SUPERSENSITIVITY FOLLOWING LESIONS

OF THE NERVOUS SYSTEM

Supersensitivity

following Lesions of

the Nervous System

AN ASPECT OF THE RELATIVITY

OF NERVOUS INTEGRATION

By

GEORGE W. STAVRAKY, M.D., C.M., M.Sc.

Professor of Physiology,
University of Western Ontario, London, Canada

With a Foreword by

WILDER PENFIELD, O.M.

Director of the Montreal Neurological Institute

University of Toronto Press

University of Toronto Press

Diamond Anniversary 1961

Foreword

BY WILDER PENFIELD, O.M.

Like the stars in the sky above us, there are millions upon millions of separate functioning cells in the nervous system of each individual. And within the microcosm that we call the brain, the actions of these cells and their reactions are controlled by laws no less obscure, no less important, than those that rule the movement of the stars.

> The heavens declare the glory of God;
> And the firmament sheweth his handiwork.

An equal claim might be made for the human brain. Many must study his handiwork there and bring to bear the varied skills of science, before the laws of action and reaction are revealed.

This book has to do with a general reaction to which neurones are subject. It is a scholarly study of the sensitization of nerve cells produced by denervation—a basic inquiry carried out by an experimental physiologist. But the work will be welcomed by clinicians just as eargerly as by physiologists, since it deals with pathological reactions of the brain. Supersensitivity may increase the action of a reflex, or augment the electrical potentials in a convolution, and this may guide the neurologist and the electroencephalographer to the correct diagnosis.

Long ago at Queen Square and in the wards of the London Hospital, Hughlings Jackson watched the working of the nervous system after disease had left its lesion in it. Comparing action with reaction, he had the wit to wonder at the principles involved. Why should a lesion in one area, or one level, of the central nervous system of man be followed by release of greater activity in another? Why should local damage to grey matter be followed by local exaltation of ganglionic activity in the very place where the nerve cell population had decreased? The explosive discharge of the remaining cells, he surmised, must be the cause of epilepsy.

Since Jackson's time, neurophysiologists have learned to measure ganglionic activity electrically. But they have, so far, failed to elucidate the reason for increased irritability in the denervated cell. They have not yet explained the exaltation of action in the epileptogenic focus, or the increase of reflexes after injury, or decerebrate rigidity, spasticity, involuntary movement, the tremor and rigidity of Parkinson's disease.

Professor Stavraky does not concern himself primarily with these clinical syndromes. But he never forgets them, knowing that his basic work may throw revealing light upon them. In a physiological laboratory he has carried out an analysis of pathological supersensitivity. He has summarized his own experimental work and assembled the relevant contributions of others in an exhaustive bibliography.

It was in the middle of the nineteenth century that Marshall Hall first pointed to an "augmentation" of functions of the nervous system that might follow initial "diminution." Later on, Claude Bernard concluded that there was augmentation of the excitability of all tissues "when separated from the nervous influences which dominate them." In recent years, Walter Cannon, in company with Rosenblueth, has described "supersensitivity of partially isolated neurones in the central nervous system as well as in peripherally innervated structures." Cannon took as the title of his Hughlings Jackson lecture at the Montreal Neurological Institute: "A Law of Denervation."(Am. J. Med. Sci. *198*: 737–750 (1939)) He pointed then to the important applications of the law. Sensitization of denervated structures, he suggested, might help to explain unsolved problems of neurology from involuntary movement to epileptic fit.

As undergraduate and graduate student, George Stavraky came early under the stimulating influence of a wise physiologist, Boris Babkin, who was one of the most distinguished of the early pupils of Ivan Pavlov. But Professor Stavraky's work is his own. It is independent and unbiased. He has devoted years of study to this subject and the book which now appears is the well-balanced, happy outcome of his analysis of physiological thought and of his own original research at the University of Western Ontario.

Preface

This book is based upon a series of lectures and talks given before the Toronto Biochemical and Biophysical Society, the Montreal Neurological Society, the Fellows Society of the Montreal Neurological Institute, and during a post-graduate course of four lectures in neurology at the University of California at Los Angeles, as well as graduate seminars in the Department of Pharmacology of the University of Michigan and in the Faculty of Medicine of the University of Western Ontario. I should like to express my thanks to my colleagues and friends who encouraged me to assemble these talks into book form and advised me in seeking its publication.

Much of the work that is discussed was done in collaboration with my co-workers and pupils in this laboratory and to them, in particular, I wish to express my indebtedness, as well as to the granting bodies who contributed to the financial support of the research. I should like to record my thanks to Dr. C. W. Gowdey for reading the manuscript and to Mrs. M. Peereboom and my wife, Madeleine Stavraky, for their help with the preparation of the latter.

Finally, I take this opportunity of thanking those publishers and editors who have given me permission to reproduce figures which appeared previously in various scientific publications as well as to make some direct quotations. The source of each figure and quotation is given in the legends to the figures and in the text.

G. W. S.

Department of Physiology
University of Western Ontario
London, Canada

Contents

SUPERSENSITIVITY FOLLOWING LESIONS

OF THE NERVOUS SYSTEM

As early as 1841, Marshall Hall recognized that "the first effect of injury done to the nervous system is a diminution of its functions; whilst the second or ulterior effect is the augmentation of those functions." In a more sweeping generalization, Claude Bernard (1880) again expressed the opinion that "the excitability of all tissues seems to augment when they are separated from the nervous influences which dominate them." These broad views of the regulation of function did not receive the full recognition they merited. Setschenow (1863, 1868, 1875) showed that inhibition of spinal reflexes could be caused by chemical stimulation of the optic lobes of the frog, and, as a clinical counterpart to this concept, the principle of the dominance in the nervous system of the higher levels over the lower ones was expounded with great success by Hughlings Jackson (1884). From his observations of epileptic seizures, Jackson (1870) developed a method of studying the localization of function in the cerebral hemispheres, which preceded the experimental studies in this field. Direct demonstration of cortical localization, which was postulated by Jackson, was provided by Fritsch and Hitzig (1870), Ferrier (1873, 1876) and others; and electrical stimulation of the cerebral cortex in human beings, carried out with improved techniques by Penfield and Rasmussen (1950), gave an insight into the details of cerebral localization in man. Although Jackson is credited with the concept of cerebral localization, Sir Victor Horsley (1907), in summarizing Jackson's contributions, expressed the belief: "The greatest principle which underlies all he has written is the true nature of localization of function in the central nervous system—namely, that it is relative and not absolute."

What is the meaning of the "relativity" of localization which Horsley emphasized as the corner-stone of Jackson's teachings? The assumption of a specificity of action of nervous connections forms the basis of many functional concepts in the nervous system. It is maintained that there are specific excitatory and inhibitory fibres or neurones which exert their action in accordance with the distribution of their terminations and that higher levels of the nervous system mediate their excitatory and inhibitory effects by means of specific anatomical connections. Many very fundamental discoveries support this view. Various nerve fibres with their differing conduction rates and other characteristics were described by Erlanger and Gasser (1937), and the functional significance of a multitude of nervous connections was studied by Adrian (1928, 1931, 1932, 1941, 1953, 1955), Matthews (1929, 1931 a, b, 1933), Lloyd (1941 a, b, 1943 a, b, 1944, 1946 a, b, 1948), and many others.

The discovery of the small fibre, or gamma-efferent, system by Leksell (1945), and the subsequent studies of Granit (1955), Hunt (1951, 1952 a), Hunt and Kuffler (1951 a, b), Kuffler and Hunt (1952), McCouch *et al.* (1950), and others, helped to clarify the role of the muscle spindles and tendon receptors in the regulation of muscular contractions. Tegmento-bulbar regulation of tone and of postural activity, which was suggested by Magnus (1924, 1926) and Rademaker (1931), was brought into a harmonious concept by Magoun, and Magoun and his co-workers (1944, 1947, 1950, 1953 a, b) and by others in their studies of the excitatory and inhibitory influence exerted by the reticular formation and by the vestibulo-spinal projections. At the same time, electrophysiological studies dealing with functional localization vastly expanded the knowledge about the significance and interrelation of different parts of the central nervous system.

The basic striving in the formulation of a general law is a desire to attain order and simplicity. However, with the progress of knowledge about detailed functional localization in the central nervous system, the concept of specificity of action of nervous connections necessitated the adoption of more and more complex patterns of organization. Furthermore, many apparent exceptions to the general rule began to emerge. Realization of this resulted in a search for other approaches to the problem of nervous integration. While Burn (1945), Babkin (1946), and Hess (1947) emphasized exceptions to the classical concept of the antagonism of the sympathetic and parasympathetic nervous system, and Loewi (1944, 1945 a, b, c), Dale (1937, 1938 a, b, c), Feldberg (1945), Spadolini (1917, 1940, 1948) and Welsh (1948, 1955), to name but a few, strove to correlate excitation and inhibition with direct humoral mechanisms, Munk (1909), Spiegel and Démétriades (1925), Bremer (1932) and Cannon (1939), revived Hall's and Bernard's idea that changes in the sensitivity of isolated neurones and peripheral structures to chemical agents and to impulses reaching them by way of remaining connections, led to exaggerated and modified effects. Cannon (1939) summed up the evidence for the occurrence of "supersensitivity" in various structures after the severance of efferent nervous connections in his "Law of Denervation." Later, in a monograph entitled "The Supersensitivity of Denervated Structures," Cannon and Rosenblueth (1949) expanded the concept almost to Bernard's original scope. They incorporated into their definition of supersensitivity the hyperirritability of partially isolated neurones in the central nervous system to impinging nerve impulses of both an excitatory and inhibitory nature as well as to chemical agents. Instances of severance of afferent connections in the central nervous system were also mentioned. Cannon's evidence for the occurrence of an exaggeration of the sensitivity in peripheral structures after their denervation is very complete and will probably remain one of the classical chapters in physiology. The evidence that he presented for the development of supersensitive states in the central nervous system following the neuronal isolation of various regions was more sketchy and left much room for further investigation and thought. Keeping this in mind, the present discussion will be centred predominantly on changes in irritability which take place in the nervous system.

The first feature which will be considered is the finding that, after partial isolation, various regions of the central nervous system can be selectively stimulated by discriminative doses of chemical agents when the latter are introduced into the general circulation of animals and human beings with chronic lesions. The co-ordinated responses thus evoked offer an opportunity to study the most complicated activity of specific portions of the central nervous system. But, what is even more important, partial isolation reproduces experimentally an unbalanced condition of the nervous system which resembles certain clinical disorders. This unbalance is reflected in the actions of various convulsant and anticonvulsant drugs and in the responses of the affected regions to nerve impulses reaching them by way of the remaining connections. Thus, the study of changes in patterns of convulsive seizures and the susceptibility to convulsions undertaken in animals with partially isolated regions of the central nervous system may be of some interest to the clinical neurologist.

In the investigations carried out in Cannon's laboratory, great emphasis was laid on the hyperirritability resulting from deefferentation. While this approach is very natural in studies which deal with the excitability of peripheral structures, in the central nervous system the importance of afferent connections made it highly desirable to have some knowledge of the behaviour of neurones isolated by means of section of the centripetal fibres. For this reason, the sensitivity of deafferented neurones in the central nervous system to chemical agents and to nerve impulses reaching them through descending paths will be considered in the second part of this treatise.

The "relative nature" of localization of function referred to by Horsley which forms a prominent feature of many investigations involving the nervous system, became most evident during the study of the responses of partially isolated regions of the central nervous system. Under optimal experimental conditions partially isolated neurones were very sensitive to stimulation. Yet a variety of factors led to opposite results: isolated neurones became depressed more easily and more profoundly than did those employed as controls. Often, excitation and inhibition seemed to be brought out by denervation in such a way that the maximal stimulation of the supersensitive structure coincided with the development of inhibition, while weaker stimulation of the same structure resulted in an excitatory response. When investigated further, this reversibility of the effects of stimulation was found to be not an exclusive property of denervated structures. It seems to be part of the general feature of excitability of living tissues and is only greatly exaggerated by denervation. The reversibility of the effects of nervous and chemical stimulation was studied both in peripheral organs and in the central nervous system, and these studies constitute the third part of this work.

Finally, it may be mentioned that throughout the present investigations the principle of sensitization by denervation was constantly put to the test, and the hyperirritability which developed after isolation was compared and contrasted, whenever possible, with the augmentation of the facilitatory influx and with the release from suppressor or inhibitory influences. For this reason, the time of onset of the augmented excitability of the isolated areas was carefully recorded and the gradual change in excitability was distinguished from the immediate effects of various operative procedures. The considerations that make the concept of sensitization by denervation plausible and advantageous when applied to certain physiological occurrences, and to some clinical derangements of the central nervous system, will be referred to as opportunities present themselves.

| **Effects of Cerebral Ablations on**

Induced Convulsions

1. GENERAL BACKGROUND

Apart from the direct bearing which studies of convulsibility have on the etiology of epilepsy, the wide adoption of shock therapy as a method of treating mental disorders has given new significance to experimental investigations dealing with the physiological background of convulsive manifestations. Mental derangements are also being treated extensively by means of surgical disconnection of the frontal lobes from the rest of the central nervous system. An understanding of the relation between convulsibility and frontal lobotomy or lobectomy, therefore, becomes one of the more urgent needs of modern psychiatry. One approach to this general problem lies in the study of the sensitivity of partially isolated regions of the central nervous system to convulsant and anticonvulsant agents.

Indications that, following denervation, peripheral structures become extremely sensitive to chemical stimulating agents can be found in the early literature. For instance, it has been known for a long time that after section and degeneration of the hypoglossal nerve, stimulation of the chorda tympani resulted in a prolonged contraction of the tongue—the Philipeaux-Vulpian (1863) phenomenon. Rogowicz (1885) noted a similar increased sensitivity of the facial muscles: after section of the facial nerve, sympathetic stimulation caused a contraction of the muscles on the denervated side. Sherrington (1894) found that stimulation of a peripheral nerve after section and degeneration of the ventral roots caused a slow contraction of the muscles of the extremities, and Bender (1938) described a contraction of previously denervated muscles in the monkey in response to fright. Subsequently, it was shown that contractions of the denervated muscles could be reproduced by acetylcholine (Frank *et al.*, 1922) and by adrenaline (Euler and Gaddum, 1931; Bülbring and Burn, 1936) and that perfusates from muscles activated by the stimulation of autonomic nerves contained an acetylcholine-like substance (Bain, 1932; Feldberg, 1933; Bülbring and Burn, 1934). It was also shown that the sensitivity of denervated muscles to acetylcholine was greater than that of muscles which had their innervation preserved (Brown and Harvey, 1938; Brown *et al.*, 1936; Kuffler, 1943; and others). This sensitivity developed gradually and became pronounced four to eight days after section of the nerves (Frank *et al.*, 1922; Knowlton and Hines, 1937; Brown, 1937; Rosenblueth and Luco, 1937; and others). Later work showed increased sensitivity of a variety of denervated structures to different drugs, but the phenomenon was demonstrated especially clearly in relation to the action of acetylcholine and adrenaline in organs which are under sympathetic or parasympathetic control.

Cannon and Rosenblueth (1936) and Rosenblueth and Cannon (1939) observed that cells in the superior cervical ganglion which had been denervated several days previously became more sensitive to acetylcholine than the cells of the opposite ganglion. A detailed description of their findings and further literature pertaining to this subject may be found in the monograph by Cannon and Rosenblueth (1949), "The Supersensitivity of Denervated Structures."

2. RESPONSES OF ANIMALS AND MAN WITH LESIONS OF THE CENTRAL NERVOUS SYSTEM TO ACETYLCHOLINE, ADRENALINE, AND METHACHOLINE

A.   *Effects of acetylcholine*

i. *Intra-arterial injections of acetylcholine in spinal animals.* The information gained during the studies on ganglionic cells and on peripheral structures suggested the possibility that denervation may also render spinal motor neurones more sensitive to chemical stimulation. This was shown by Cannon and Haimovici (1939) to be the case in cats in which semisection of the spinal cord was carried out. At various intervals of time after the operation, high-spinal transections were made, and the contractions, induced by intra-aortal injections of acetylcholine, strychnine, sodium carbonate, and by asphyxia, were recorded with isotonic levers. The quadriceps on the side which was previously semisected responded to smaller quantities of the stimulating agents, and the contractions were more pronounced than those on the control side.

They found that semisection of the spinal cord also increased the sensitivity to acetylcholine of the quadriceps muscle itself. This effect of

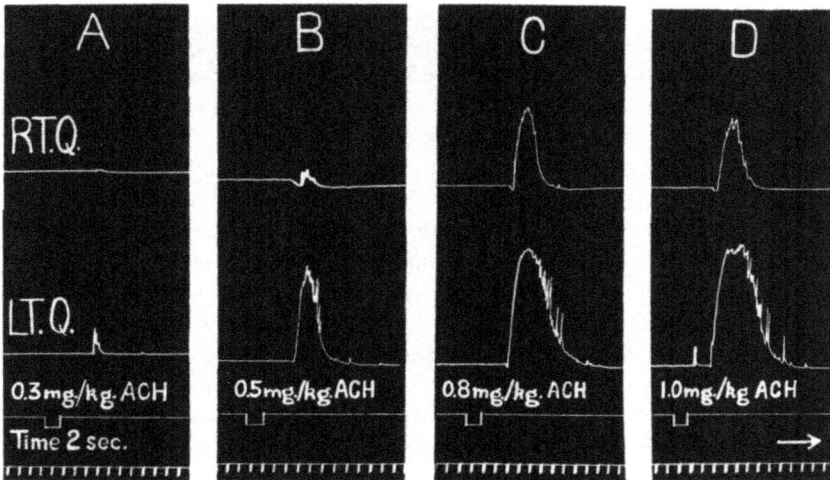

FIG. 1. Isotonic myograms showing the effects of increasing quantities of acetylcholine in a cat nine days after left spinal semisection at T 12 level (RT.Q., right quadriceps, LT.Q. left quadriceps). Acetylcholine was injected intra-aortally in a high-spinal preparation. Note lower threshold and greater effects on the side of the semisection. (Drake, Seguin and Stavraky, 1956)

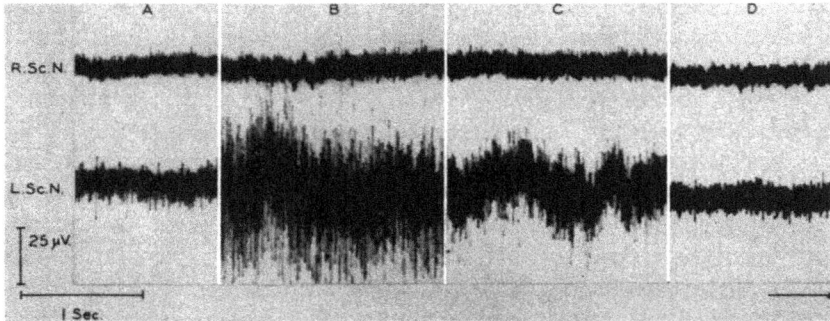

FIG. 2. Effects of intra-aortal injection of 0.09 mgm./100 gm. of acetylcholine in a high-spinal white rat twenty-five days after left semisection of the cord at T 12 level. Oscilloscopic recordings from the central ends of severed sciatic nerves (R.Sc.N., right sciatic nerve, L.Sc.N., left sciatic nerve). *A*: activity three seconds after the start of the injection (injection time two seconds). Note beginning of increase in activity of the left side but normal activity on the right. *B*: activity nine seconds after start of the injection—maximal effect of acetylcholine. *C*: activity sixteen seconds after start of the injection is subsiding, but still increased on both sides. *D*: thirty-nine seconds after start of the injection, activity on the right side has returned to the pre-injection level, while that on the left side is still somewhat increased. (Drake, Seguin and Stavraky, 1956)

"decentralization" of the muscle persisted after section of the motor nerves, and was referred to as "penultimate sensitization."*

These findings were substantiated and expanded by Drake (1947), Drake and Stavraky (1948 a) and by Drake, Seguin and Stavraky (1956). Using a technique based on the one employed by Cannon and Haimovici (1939), it was found that four to sixty-four days after semisection of the spinal cord doses of acetylcholine ranging from 0.1 to 1.0 mgm./kgm., injected through a plastic catheter inserted into the aorta by way of the subclavian or external carotid artery, had greater effects on the side of the semisection, not only in conditions of myographic recording, but also when the peripheral nerves were severed and the electrical activity of the anterior horn cells was studied. Thus it was demonstrated that the augmented response of the quadriceps on the semisected side was due to an increased number of nerve impulses coming from the previously isolated area of the spinal cord rather than to an increased sensitivity of the muscle itself. (Figs. 1 and 2)

Finally, it was shown that a greater sensitivity of the spinal neurones to intra-arterial injections of acetylcholine and to other chemical agents developed not only following a semi-section of the cord, but also after a semidecerebration, or when one frontal lobe including the motor cortex was extirpated. These experiments will be discussed more fully in the concluding section (pages 123–126, Figs. 41 and 42), but they are mentioned here because they eliminate the possibility that unequal recovery from the shock of transection on the two sides of the cord may have influenced the results. If this view were taken, it could be argued that the motor neurones on one side were protected by the preceding semisection from the consequences of the complete transection of the cord carried out at the time of the acute experiment and thus had time to recover from the effects of the isolation. This objection was overcome by placing the initial one-sided lesion *above* the subsequent spinal transection of the cord.

All these experiments proved conclusively that acetylcholine stimulated the isolated spinal neurones to a greater degree than it did those on the opposite side of the spinal cord. This greater sensitivity developed gradually and was evident three to four days after semisection. In the experiments with cerebral ablations the asymmetry did not begin to develop until some three weeks after the operation and became pronounced some three months after the operation. This gradual onset of the hyperexcitable state of the neurones offers a most convincing proof

*The terms denervation and decentralization are used as defined by Cannon: "denervation" referring to section of direct neuronal connections, and "decentralization" to the severance of once-removed or "penultimate" neurones.

that a true change in sensitivity to chemical stimulating agents is brought on by the neuronal isolation.

ii. *Intravenous injections of acetylcholine in cats after removal of a frontal lobe or of a cerebral hemisphere.* The possibility that intravenous injections of acetylcholine might also exert selective effects on previously denervated regions of the central nervous system was investigated by Stavraky (1943). In cats, cerebral ablations were carried out ranging from the removal of one frontal lobe to a semi- or even to a complete supratentorial decerebration, the motor cortex always being included in the ablations. After recovery, acetylcholine was injected, without anaesthesia or preparation of any kind, into the femoral or great saphenous vein once a week at first, but later at longer intervals.

In intact cats, 0.1–0.2 mgm. of acetylcholine caused only slight activity: a transient dilatation of the pupils, salivation, and a brief phase of unsteadiness or some vermiform movements of an athetoid type of the extremities. When the dose was increased, acetylcholine (0.3–0.6 mgm.) sometimes produced slow spastic contractions of the extremities which terminated in a generalized tonic convulsion of an extensor type, the head being slowly drawn backwards on the neck and the extremities thrust out in spasm. This was followed occasionally by some clonus,

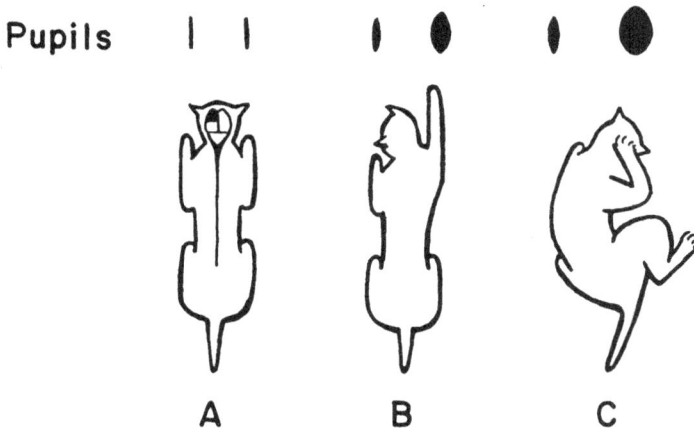

Fig. 3. Diagrammatic representation of the responses to I.V. injections of subconvulsant doses of acetylcholine in a cat from which the left frontal lobe, including the motor cortex, was removed several months previously. *A*: pupils and posture before the injections. *B*: quick turning of the head to the left, and a forward thrust of the right front paw, preceded by an unequal dilatation of the pupils; seen after 0.05 to 0.1 mgm. of acetylcholine. *C*: contraction of the whole right side of the body and unequal dilatation of the pupils; after 0.1 to 0.2 mgm. acetylcholine. (Stavraky, 1943)

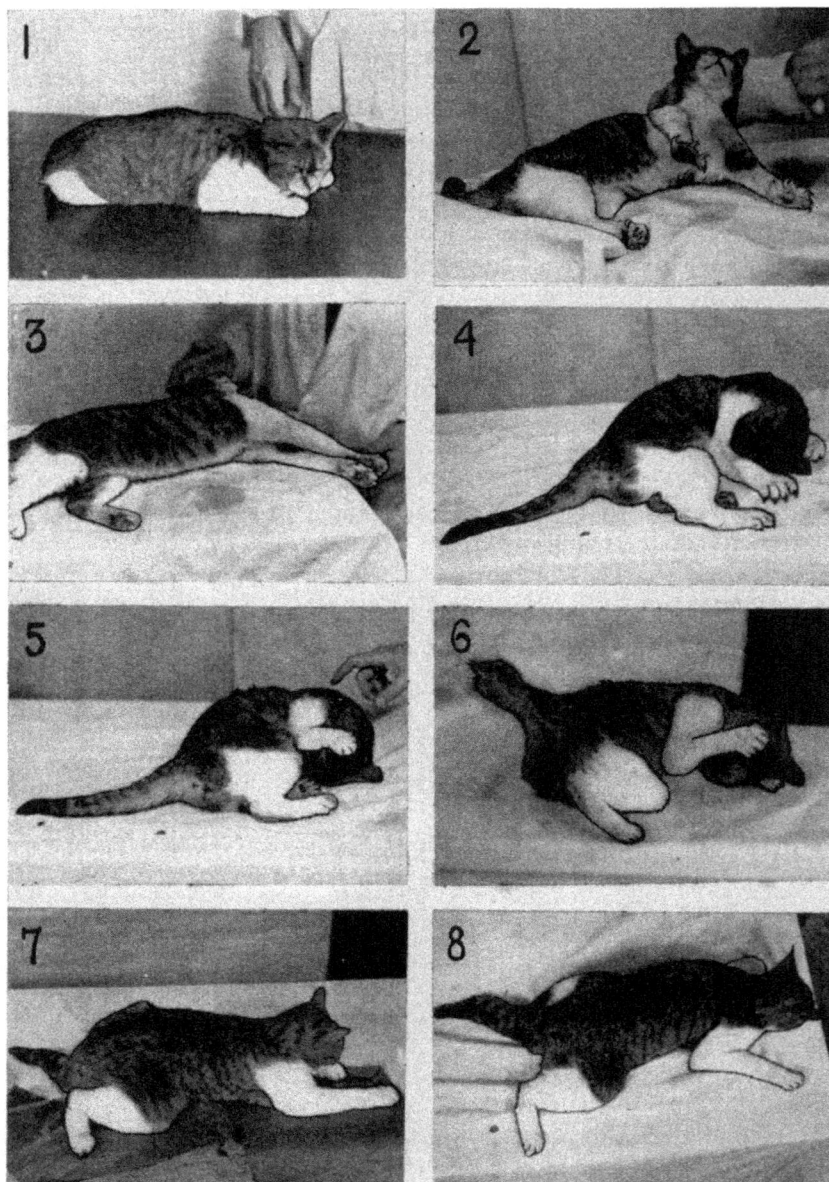

FIG. 4. Effects of intravenous injection of 0.25 mgm. of acetylcholine in a cat in which the left frontal lobe, including the motor cortex, was removed four months previously. 1: posture before injection. 2: generalized tonic convulsion. 3: phase which resembles transient decerebrate rigidity. 4, 5, 6: slow secondary contraction in flexion, of the *right side only*. 7, 8: extensor hypertonicity replaces

dilatation of the pupils, salivation, lacrimation, audible gastro-intestinal motility, and sometimes by micturition and defecation. Such seizures were followed by a period of depression during which the animal lay on its side with flaccid limbs, contracted pupils, and stertorous breathing.

When acetylcholine was injected into the operated cats, these animals showed all the signs of a selective excitation on the side of the body opposite to the removal. With repeated weekly injections this asymmetrical response was found to reach a maximum in five or six weeks after the operation and persisted thereafter. Small quantities of acetylcholine (0.05–0.1 mgm.), which were almost ineffective before the operation, now produced a forward thrust of the contralateral front paw and a turning of the head towards the side of the operation. These motor manifestations were preceded by a transient dilatation of the pupils which was more pronounced on the side opposite to the cerebral ablation (Fig. 3 B).

Injections of 0.1 to 0.2 mgm. of acetylcholine resulted in a greater dilatation of the contralateral pupil and a contraction of the entire side of the body opposite to the removal. The head was drawn down and the face turned away by a contraction of the sternocleidomastoid muscle; the front paw was raised over the head and often the protruding claws were pressed into the skin behind the ear. The back paw was drawn forward in a semiflexed position, and the body of the animal was curved with the concavity away from the side of the operation (Fig. C). Often there was some pilo-erection over the spastically contracted extremities. With repetition of the injections, the duration of these tonic contractions gradually increased and in some animals lasted as long as four or five minutes. Towards the end of the contraction a coarse tremor occasionally appeared in the rigid limbs. When still larger doses of acetylcholine were injected (0.2–0.3 mgm.), a bilateral tonic stiffening of the body developed. It had the appearance of an extensor spasm, and when the animal came out of this rigid state, the ipsilateral limbs quickly acquired their normal posture and regained their natural tone, while the opposite side of the body assumed the characteristic curvature. Sometimes vermiform movements of the extremities developed on the operated side simultaneously with a tonic stiffening of the opposite side of the body. The various stages of such a generalized convulsion, induced by acetylcholine in a cat from which the left frontal lobe was removed four months previously, are illustrated in a series of photographs in Fig. 4.

Following the removal of both frontal lobes in one operation, the

flexion on the right side and is followed by awkwardness and inco-ordination on that side. The complete sequence lasted seven and a half minutes. (Stavraky, 1943)

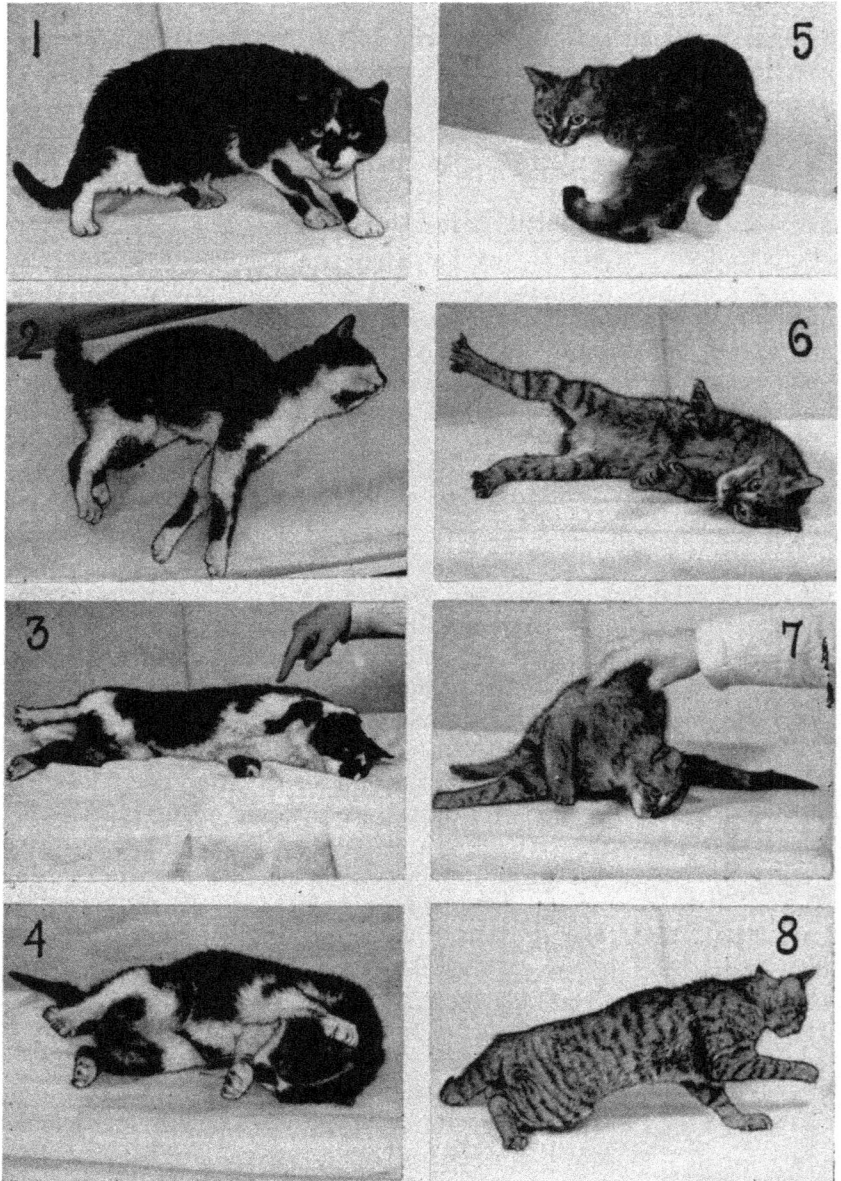

FIG. 5. 1, 2, 3, and 4 show effects of an injection of 0.3 mgm. of acetylcholine (I.V.) in a cat in which the left frontal lobe was removed four months previously, and the right frontal lobe was removed three weeks before the injection. 1:

response to acetylcholine was symmetrical on the two sides of the body. However, acetylcholine seemed to produce an exaggeration and prolongation of the tonic muscular contraction and a great accentuation of all the features of sympathetic discharge. Maximal and prolonged dilatation occurred in both pupils, the back was arched, the head retracted, the four extremities rigidly extended, the tail dorsiflexed, and the hair stood erect all over the body. When the two frontal lobes were removed in two separate operations, sometimes spaced widely apart, the gradual onset of the sensitivity on the corresponding sides was very distinct. Thus in one animal, from which the left frontal lobe was removed three and a half months before the right one, during the three weeks following the second operation 0.3 mgm. of acetylcholine caused the right front paw to be flexed and raised over the head, as shown in Fig. 5(1–4), while the left paw was held rigidly extended. In subsequent trials the left paw also began to be flexed and raised over the head, and after several more weekly injections the difference in the response of the two sides disappeared.

After semidecerebration the animals were particularly sensitive to acetylcholine and responded even to relatively small quantities of this agent by severe generalized convulsions. In these animals too, acetylcholine caused the contralateral pupil to dilate more widely than the one on the side of the operation. Towards the end of the convulsive seizure contractions of the muscles developed in the extremities only on the side opposite to the cerebral ablation. They were often rhythmic, and the front paw was flexed or worked up and down, while the hind limb was held rigid in a semiflexed position until it began a movement suggesting an incomplete scratch reflex (Fig. 5 (5–8)).

Appropriate doses of acetylcholine thus produced co-ordinated muscular contractions which were limited to that side of the body opposite the one from which the frontal lobe or a cerebral hemisphere had been removed. Larger doses caused more persistent effects on that side, the ani-

---

posture before injection. 2: generalized tonic convulsion with arching of the back, maximum dilatation of the pupils, and erection of hair. 3 and 4: right front paw is raised over the head but not the left one. Effects of the injection lasted six and a half minutes.

5, 6, 7 and 8 show effects of an injection of 0.1 mgm. of acetylcholine (I.V.) in a semidecerebrated cat three and a half months after the operation (left cerebral hemisphere removed). 5: posture before injection (note circling to the left). 6: generalized tonic convulsion. 7: end of convulsion and beginning of flexion of right front paw. 8: cat stands and walks with rigid fore limb held in a typical position; adduction at the shoulder, flexion at the elbow and extension at the wrist; hind limb is stiffened in a semiflexed position. Effects of injection lasted four and a half minutes. (Stavraky, 1943)

mals showing not only a lowered threshold to acetylcholine but a more prolonged after-effect. Another aspect of the sensitization was a greater discharge of central sympathetic neurones. This was particularly prominent when both frontal lobes were removed. Whereas in normal animals intravenous injections of 0.2–0.3 mgm. of acetylcholine evoked only a slight dilatation of the pupils, followed by constriction, the same doses after frontal lobectomy caused maximal and persistent pupillary dilatation and pilo-erection. In unilateral ablations these effects were more pronounced on the denervated side.

B.   *Effects of adrenaline*

An interesting outcome of this study was the finding that adrenaline administered intravenously in doses of 0.3 mgm./kgm. or less to semi-decerebrated or unilaterally frontal-lobectomized cats caused, after a transient inhibition, responses not unlike those elicited by acetylcholine: a predominant dilatation of the pupil and some extensor rigidity in the extremities contralateral to the operation (Fig. 6) (Stavraky, 1947).

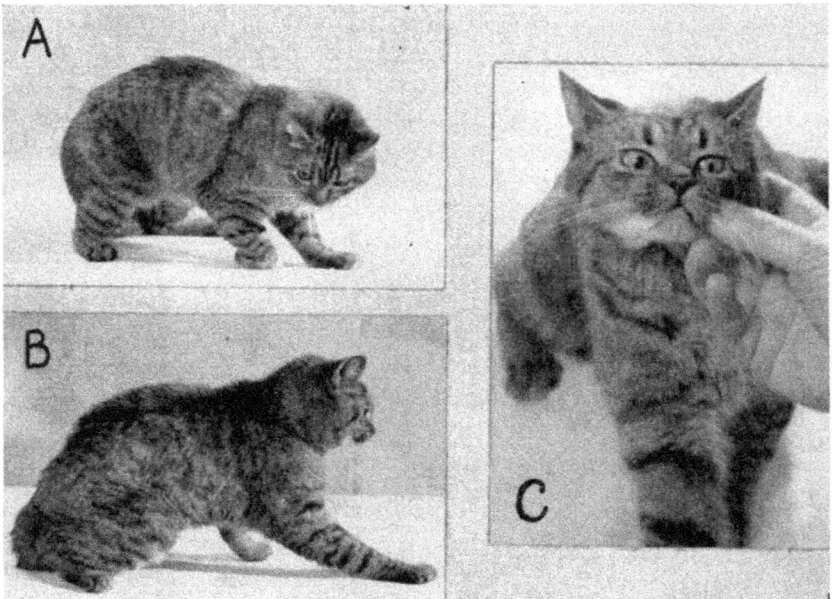

FIG. 6. Effects of adrenaline (0.03 mgm./kgm. I. V.) in a cat three years after the removal of the left frontal lobe (including the motor cortex). Initial weakness of right extremities (*A*) is followed by extensor hypertonicity (*B*). *C* shows forward thrust of rigidly extended right forepaw and predominant dilatation of the right pupil at the height of the effect. (Stavraky, 1947)

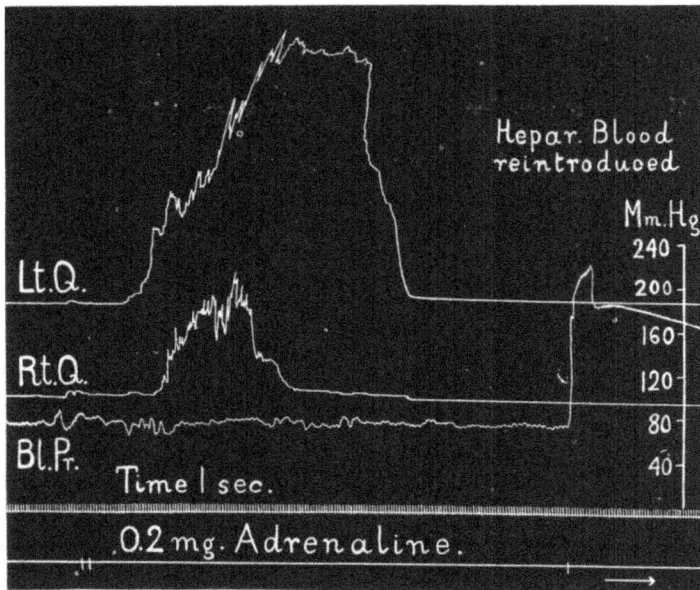

Fig. 7. Effects of adrenaline in a spinal cat twenty-one days after left semi-section of the cord. Intra-aortal injection of crystalline adrenaline causes marked contraction of the left quadriceps (Lt.Q.) which is followed by a weaker contraction of the right quadriceps (Rt.Q.). The blood pressure (Bl. Pr.) is stabilized by means of a mercury compensator. The effects of the next two injections were less pronounced and subsequent injections were ineffective. Injections with un-controlled blood pressure produced similar results. (Stavraky, 1947)

This effect of adrenaline was further investigated in acute experiments on spinal cats following chronic semisection of the cord. The iso-lated quadriceps on the side of the semisection responded to intra-aortal injections of adrenaline by a marked tonic contraction. This contraction was most pronounced after the first injection of adrenaline and was sometimes preceded by a period of relaxation of the muscle. The second and third injections evoked smaller responses, and subsequent injections were usually ineffective. In sensitive preparations the quadriceps on the control side also responded with a contraction which came on later, was much smaller, and lasted a shorter time than that on the denervated side.

As shown in Fig. 7, these effects of adrenaline did not depend on the rise of the systemic blood pressure, but they could not be readily accounted for, because Marrazzi (1939 a, b, 1953), Lundberg (1952), and Paton and Thompson (1953) have found that adrenaline had a depressant action on synaptic transmission in the nervous system. Ivy and co-workers (1944), and Leimdorfer and Metzner (1949) induced anaesthesia by adrenaline, and Feldberg and Sherwood (1954) described areflexia after intraventricular injections of adrenaline; while Gellhorn et al. (1939) showed that adrenaline in small doses inhibited or diminished the severity of pentylenetetrazol-induced convulsions. On the other hand, the recent electroencephalographic study by Minz and Domino (1953) confirmed earlier observations by Notkin and Pike (1931), Keith (1933), Keith and Stavraky (1935) and Hall (1938) that adrenaline potentiated experimentally induced convulsive seizures. The findings of Bülbring et al. (1948) and of Bernhard and Skoglund (1953) that adrenaline markedly enhanced extensor movements and monosynaptic reflexes in the spinal cord were supported by Siggs, Ochs and Gerard (1955), who found that cortically evoked leg movements and the patellar reflex may be enhanced by adrenaline. These findings and the experiments of Bülbring and Burn (1941, 1942), Bülbring (1944), and Bülbring and Whitteridge (1941), in which it was demonstrated that adrenaline potentiated the central effects of acetylcholine, enhanced transmission in sympathetic ganglia, and increased the height of action potentials evoked by submaximal stimuli in a nerve, favour the general conclusion that adrenaline may exert an excitatory effect on neurones of the central nervous system in certain experimental conditions. It has also been suggested that adrenergic neurones are present in the reticular formation which can be activated by adrenaline and thus indirectly influence remote parts of the central nervous system (Dell et al., 1954; Bonvallet et al., 1954; Vogt, 1954, 1957). Once the excitatory action of adrenaline is conceded, it is not surprising that the effect becomes particularly prominent when these neurones have been sensitized by partial isolation.

C.   *Effects of methacholine in human subjects*

During the Second World War Fisher and Stavraky (1944) had under their observation eleven male patients with lesions of the rostral portion of one cerebral hemisphere with and without involvement of the upper motor neurones. The various lesions were: two brain tumours, one frontal lobectomy for a tumour, seven head injuries in which damage was localized to the frontal lobe with or without extension of the injury to the motor cortex, and one case of a lesion following an attack of cerebro-spinal meningitis with residual signs of unilateral involvement of the pyramidal system. As controls, six normal persons, one patient with Raynaud's disease and one case of injury to the spinal cord, were

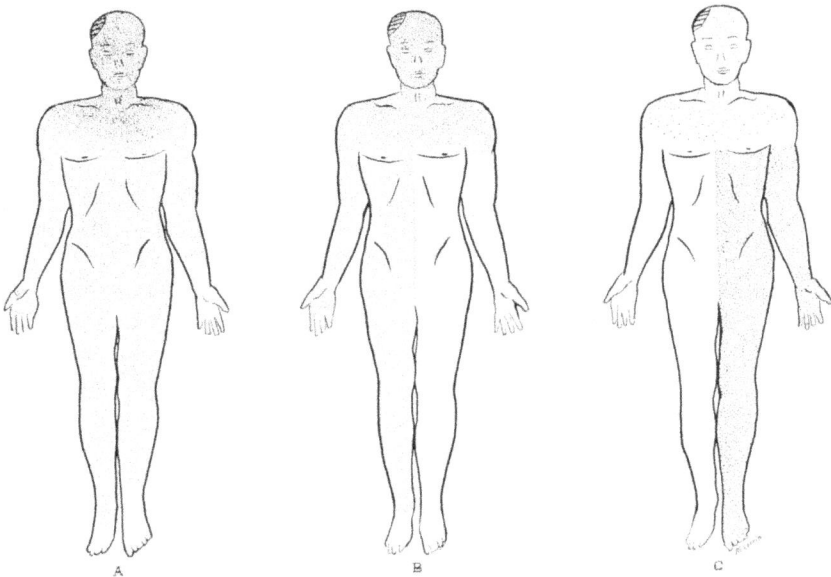

FIG. 8. Distribution of the flush on intramuscular injection of 8 mgm. of methacholine in a patient with a left-sided hemiplegia suffered as the result of a gunshot wound of the head. The bullet travelled from below up, destroying almost exclusively the motor cortex of the right cerebral hemisphere. *A*: distribution of the flush one to two minutes after the injection. *B*: distribution of the flush three and a half minutes after the injection. Note the pallor over the left side of the body in contrast to the flush on the right side. *C*: initial flush recedes about four minutes after the injection and a secondary flush appears over the previously pale regions of the left side. (Fisher and Stavraky, 1944)

studied. Methacholine (Mecholyl—Merck) was injected in quantities ranging from 8 to 25 mgm. into the deltoid muscle on two or more days.

i. *Normal males*. Though the sensitivity of individual subjects varied considerably, the tested individual usually experienced twenty to thirty seconds after the injection a sensation of heat in the face. This was followed by a flushing of the face and neck, the erythema spreading rapidly to the chest and back and then to the extremities, and decreasing in intensity as it spread to the lower part of the body. The flushing reached its maximum about two minutes after the injection and was accompanied by widespread diaphoresis and often by rhinorrhea, salivation, and lacrimation. Shivering usually occurred later in the response, and a slight increase of tendon jerks was occasionally noted. In two very sensitive individuals the first signs of flushing were accompanied by a slight transient dilatation of the pupils. Most of the features of the reaction were terminated within ten minutes of the injection.

ii.   *Patients with lesions of the rostral portions of the cerebral hemispheres.* In patients with lesions of the frontal lobe the injection of methacholine produced an asymmetrical response. In nine patients the erythema spread more rapidly on the side of the lesion and stopped at the wrist or at the metacarpo-phalangeal joints, the hand or fingers becoming very pale. In the lower extremity the flush extended to the ankle or to the dorsum of the foot. Contralaterally, the spreading of the flush was delayed and it stopped high on the limbs. Frequently, the arm, from the elbow down, and the entire leg became pale; this pallor of the contralateral extremities developed more slowly than the ipsilateral flush. It became pronounced two minutes after the injection and reached its greatest intensity about ninety seconds later, gradually replacing the initial erythema. In three patients the pallor spread over almost the whole side of the body opposite to the lesion, only the chest above the nipple line and the face remaining flushed on that side. About four minutes after the injection the contralateral blanching was followed by a flush; this occurred after the general redness had largely subsided (Fig. 8). During the reaction the contralateral extremities were markedly colder than the ipsilateral ones. There was also an asymmetrical distribution of sweating, which was less pronounced on the side of the body opposite the lesion. In the early stages of the response a delay in the sensation of heat on the opposite side of the face and a transient dilatation of the contralateral pupil were noted in seven patients. In four of them, both pupils dilated slightly, but the contralateral effect was more marked. On three occasions the contralateral pupil became irregular in outline and did not contract as readily to light as did the ipsilateral one.

In cases in which the lesions extended to the motor or pre-motor cortex, muscular tremors, slight involuntary movements, an increase in spasticity, pronounced hyperreflexia, clonus of the wrist, ankle and patella, and prominence of pathologic reflexes on the opposite side of the body characterized the later stages of the reaction. The blanching of the extremities was most noticeable after the injection of 8 to 10 mgm. of methacholine, but the other features were more marked when larger quantities were administered. All these phenomena were most pronounced in the case of lesions involving the motor and pre-motor areas, but a considerable asymmetry of the response was observed in frontally situated injuries in which no objective neurologic signs were present.

Normally methacholine has a predominantly parasympathomimetic action and its effects are exerted largely on the peripheral effectors. It appears from the observations recorded above that, in accordance with Cannon's Law of Denervation (1939), damage to the motor and pre-motor cortex, as well as to the cortical representation of the sympathetic nervous system, results in a sensitization of the next subordinate links in the chains of descending nervous connections. Consequently, the sensitized sympathetic neurones are stimulated by methacholine to such a degree that the usual parasympathomimetic effects of this drug are superseded by the discharge of nerve impulses originating in these neurones. The phenomena observed in the human cases could be accounted for by this mechanism. The initial flush probably resulted from vasodilatation due to the peripheral effects of methacholine; then vasoconstriction supervened because the drug stimulated the sensitized sympathetic centre contralaterally to the lesion. Finally, when the dominant vasoconstriction imposed from the centre subsided, there was a secondary flush which was possibly caused by the persistent peripheral effect of methacholine or to a reactive hyperaemia. The movements which occurred in the paralysed limb are explicable, as in the experiments on cats, in terms of sensitization of co-ordinating motor centres.

3. RESPONSES OF ANIMALS WITH LESIONS OF THE CENTRAL NERVOUS SYSTEM TO PENTYLENETETRAZOL, CAMPHOR, AND OTHER ANALEPTICS

If the conclusion reached in the preceding section is correct, then stimulation of the central nervous system by means of chemical convulsant agents such as absinthe, camphor, or pentylenetetrazol should yield essentially similar results. Yet, the lack of uniformity of opinion on this matter is quite surprising. Sparks (1927), Bertha (1928), Pike and Elsberg (1925), Uyematsu and Cobb (1922), Hahn (1941 a, b, c), van Bork, Dalderup and Hirschel (1950), Asuad (1940), Ten Cate and Swijgman (1945), and many others, using different techniques and criteria, came to the conclusion that cerebral ablations diminished the sensitivity of animals to convulsant agents, while Sauerbruch (1913), Dandy and Elman (1925), and Dandy (1927) maintained that the reverse was true. This latter view is supported by van Harreveld's observation (1947) that decerebrate rigidity was enhanced by "electroshock" over longer periods of time than the duration of a tonic convulsion induced by a similar procedure in intact animals. In most of these investigations adequate consideration was not given to the time for recovery from the operation and too few animals were used for a statistical evaluation of the results. Particularly controversial were the findings of Pike and Elsberg (1925) on the one hand and those of Dandy and Elman (1925) on the other: the former suggested that cerebral ablation caused a decreased susceptibility to convulsions whereas the latter maintained that the susceptibility was increased. Pike, Elsberg, McCulloch and Rizzolo, in a later publication (1929), did seem to find a greater sensitivity to convulsants in animals with a bilateral excision of the motor cortex after prolonged recovery, but their conclusions were guarded because only four cats were used. With the availability of cinematographic recordings and careful experimental design it seemed that this discrepancy might be accounted for. Drake, Seguin and Stavraky (1956) undertook such an analysis.

A. *Intra-arterial injections of pentylenetetrazol, camphor, picrotoxin, and strychnine in spinal animals with preceding semisection of the spinal cord*

Unlike strychnine or acetylcholine, which affect all the levels of the central nervous system, agents such as pentylenetetrazol and camphor are usually considered to exert their action predominantly on the most highly developed parts of the brain. When sufficiently large doses of these drugs are employed, however, convulsions are also produced in spinal and decerebrate animals. This was particularly true of experiments in which camphor or pentylenetetrazol were injected intra-aortally. Employing a technique similar to that used in the investigation of the action of acetylcholine, Drake, Seguin and Stavraky (1956) recorded, bilaterally, the contractions of the quadriceps muscles and the electrical activity of the anterior horn cells in high-spinal cats and Wistar rats.

Pentylenetetrazol (Metrazol—Bilhuber-Knoll), injected in doses of 8 to 10 mgm./kgm. caused selective contractions of the quadriceps on the side of the semisection with little or no effect on the control side. Larger quantities of pentylenetetrazol (10–30 mgm./kgm.) caused a series of marked bilateral contractions of the muscles which started earlier on the side of the preceding semisection and lasted longer (Fig. 11A). Very large quantities of pentylenetetrazol were occasionally seen to cause an initial relaxation of the muscles which lasted four to five seconds and was more pronounced on the semisected side. This relaxation was succeeded by a series of bilateral contractions. Electrical recordings from severed femoral or sciatic nerves showed that the drug caused a markedly greater activation of the anterior horn cells on the semisected side (Fig. 9). In Wistar rats, 10 to 36 days after semisection of the cord, pentylenetetrazol (2–20 mgm./kgm.) produced a mean increase in electrical activity of the spinal motor neurones on the intact side of 13 micro-volts, while on the side of the semisection the increase was 22 micro-volts. The difference between these values was statistically significant.

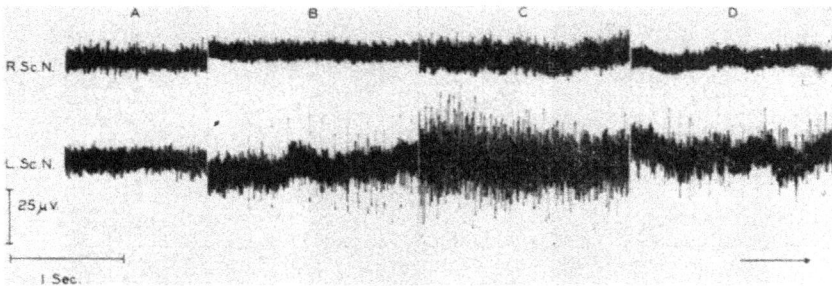

FIG. 9 Effects of an intra-aortal injection of 0.23 mgm./100 gm. of pentylene-tetrazol in a high-spinal white rat 25 days after left semisection of the cord at T 12 level. Oscilloscopic recordings from the central ends of severed sciatic nerves (abbreviations as in Fig. 2). *A*: control taken one hour before the injection. *B*: twenty-one seconds after the start of the injection (injection time seven and a half seconds). Note increase in activity on the left side but not on the right. *C*: fifty-five seconds after the start of the injection—maximal effect. *D*: seventy-five seconds after the start of the injection, the effect persists on the left side, while on the right side the activity returned practically to a normal level. (Drake, Seguin and Stavraky, 1956)

Camphor, picrotoxin and *strychnine* were also injected on numerous occasions and exerted more marked effects on the side of the semisection (Fig. 10). It is interesting that Claude Bernard (1880) drew attention to the fact that strychnine can selectively cause movements in the paralysed limbs of paraplegic patients. Fisher and Stavraky (1944) saw a similar effect with methacholine in a patient with a partial paralysis of a quadriplegic type following injury to the spinal cord at the level of the cervical 6 to thoracic 1 segments brought about by a fracture dislocation of the spine.

    i. *Relation between circulatory and convulsant actions of the drugs.* Hall (1938), Hahn (1941 a, b, c) and others found that the convulsant action of various drugs was associated with a rise of systemic blood pressure. This rise was minimized in the experiments of Drake, Seguin and Stavraky (1956) by the technique of intra-arterial injections. In spite of this the electrical discharges from the anterior horn cells and the muscular contractions were quite prominent (Figs. 7 and 11).

B. *Convulsions induced by pentylenetetrazol and camphor in cats after removal of one motor cortex, one frontal lobe or after semi-decerebration*

    A study of the effect of convulsant agents on animals with various lesions of the central nervous system gave similar results.

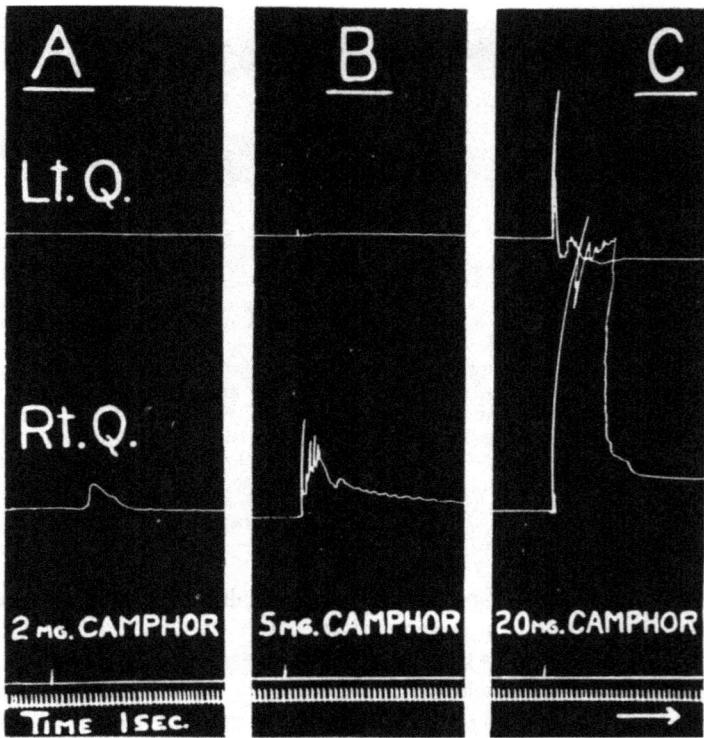

Fig. 10. Isotonic myograms showing the effects of increasing quantities of camphor monobromate injected intra-aortally into a high-spinal cat (2.3 kgm.) ten days after right semisection of the cord at T 12 level. Abbreviations as in Fig. 7. Note lower threshold, shorter latency and greater and more prolonged contractions of the quadriceps on the previously semisected side. (Drake, Seguin and Stavraky, 1956)

   i.   *Pentylenetetrazol.* Two separate experiments with pentylenetetrazol were carried out. In the first one (Drake, 1947; Drake and Stavraky, 1948 a), the first injections were made forty-eight to seventy-two hours after a unilateral ablation of the sigmoid gyrus of a frontal lobe or after a semidecerebration and were continued over periods up to seven months. The patterns of convulsions and effects of repeated injections upon them as well as the optimal spacing of the injections were studied, and moving pictures and other photographic recordings of typical convulsions were made. In this series acetylcholine, pentylenetetrazol, or camphor were injected in a random fashion and were repeated over prolonged periods

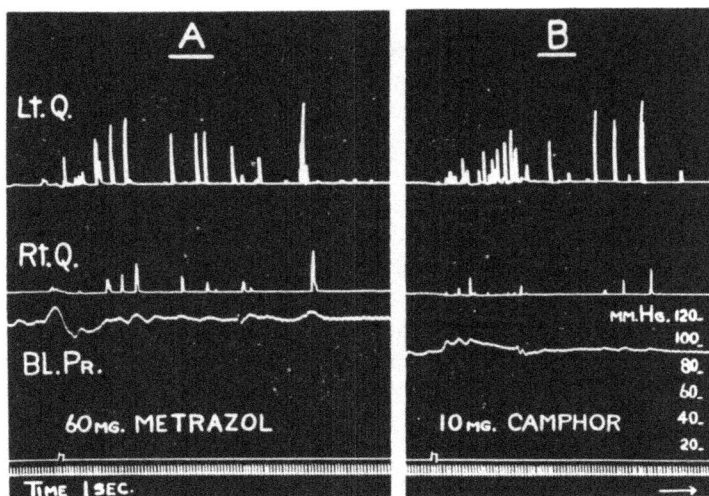

FIG. 11. Isotonic myograms showing the effects of intra-aortal injections of pentylenetetrazol and camphor in a high-spinal cat (2.4 kgm.) eight days after left semisection of the spinal cord at T 12 level. Blood pressure stabilized by means of a mercury compensator; abbreviations as in Fig. 7. (Drake, Seguin and Stavraky, 1956)

of time. The animals seemed to develop a particular proneness to asymmetrical convulsions, which became greatly exaggerated towards the end of the study. For this reason an experimental design which allowed for a statistical analysis of convulsibility was used in a new group of cats which were given a maximum of three or four consecutive weekly injections of pentylenetetrazol (Drake, Seguin and Stavraky, 1956).

In this experiment the mean durations of the studied components of the convulsions were determined, the differences between groups tested for significance, and the median convulsant dose for pentylenetetrazol was calculated.

(a) *Description of convulsions.* In *intact cats*, 6, 8, 10, and 12 mgm./kgm. pentylenetetrazol caused one of the following three types of response: (1) a mild convulsion which manifested itself in an excitation followed by twitching of the ears, blinking, and several generalized twitches of the extremities; (2) a pronounced clonic convulsion (CL) which started with flexion of the limbs, the animal falling to the ground and having a series of rapid extensor twitches; or (3) a clonic-tonic-clonic (CTC) type of convulsion which started in the same way as the previous one, but in which the initial clonic phase was followed by a period of generalized tonic flexion. Towards the end of the tonic period the hind limbs were gradually extended and began to twitch; this marked the onset of the terminal clonus which spread to the rest of the body and ended abruptly, the animal developing a transient opisthotonos and extension of all four limbs. The convulsions were followed by a period of flaccidity which was sometimes interrupted by running movements and secondary twitches. Autonomic manifestations such as transient dilatation of the pupils followed by constriction, micturition, defecation and pilo-erection, as well as vocalization, accompanied the convulsions.

The *operated cats* responded to pentylenetetrazol by a modified clonic or CTC-type of convulsion. These convulsions were asymmetrical in character and their pattern depended greatly on the duration of recovery from the operation, previous injections, quantity of convulsant agent employed, and other variables. Forty-eight to seventy-two hours after the removal of a left motor cortex, frontal lobe or cerebral hemisphere, the convulsion had some clonic and tonic components on the left side, but consisted predominantly of an extensor tone on the right side of the body. With prolonged recovery after the operation 4, 6, 8, and 10 mgm./kgm. of pentylenetetrazol caused another type of asymmetry, which could be grouped into four types of response depending upon their severity. (1) A minimal convulsion during which either the cat circled rapidly away from the side of the operation or unilateral clonic extensor twitches occurred in the right extremities. (2) The animal fell and rolled to the right; the left limbs were in clonus but the right front limb was tonically flexed (Fig. 12 A, B). Following this, the clonus gradually involved the right limbs and soon became more violent on the right side than on the left (Fig. 12 D). (3) The sequence was the same but a generalized tonic phase came into prominence (Fig. 12 C). (4) A bilateral flexor tonus was followed by a transient clonus and then by a second tonic phase. The first two convulsive patterns could be considered as belonging to the clonic type seen in intact animals, whereas the last two were comparable to the clonic-tonic-clonic responses observed in the controls. During recovery from the convulsions, after a short period of flaccidity and areflexia, there was hypertonicity of the right limbs (Fig. 12 F), forced grasping movements of the right forepaw and the animal circled and leaned to the side of the lesion.

In a few of the animals secondary convulsions were seen. These occurred occasionally only on the right side, the animal walking normally on the left extremities while the right ones were clonically convulsing. The impression was gained that the severity of the autonomic manifestations—for example, dilatation of the pupils, defecation, micturition and

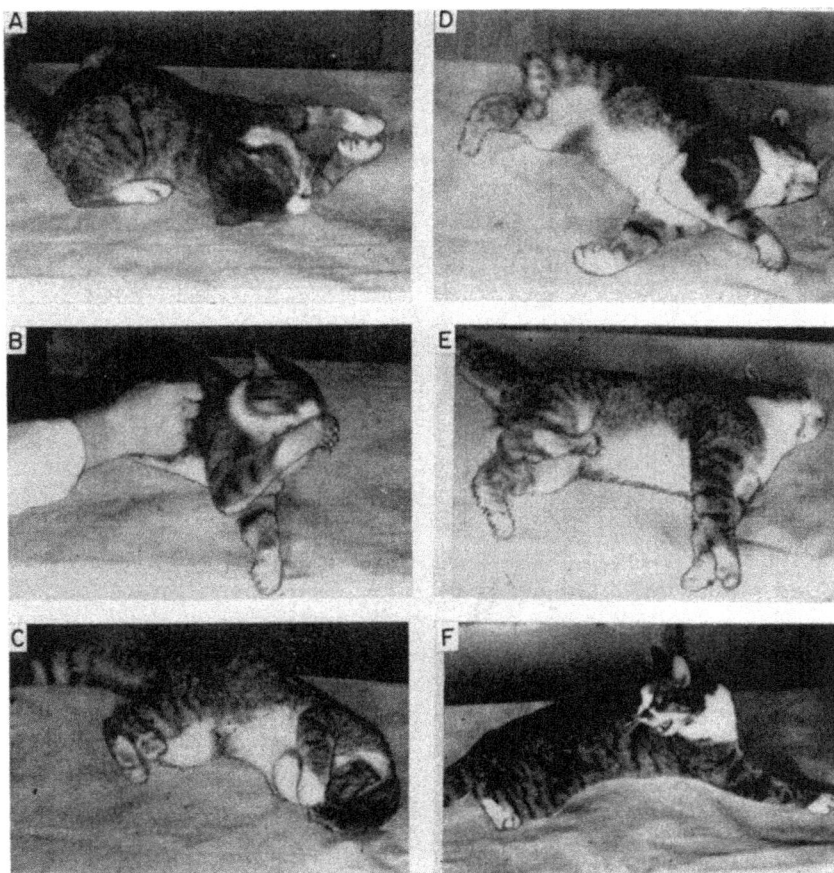

Fig. 12. Effects of an intravenous injection of 8 mgm./kgm. of pentylenetetrazol in a cat one year after the removal of the left frontal lobe. *A*: rolling to the right (note tonic flexion of right fore limb and protruding claws). *B*: end of rolling and beginning of the tonic phase. *C*: tonic phase (in flexion). *D*: clonic phase (note predominance of activity in the right limbs). *E*: terminal extension. *F*: post-convulsive hypertonicity of the right side. (Drake, Seguin and Stavraky, 1956)

pilo-erection—was greater, and that these manifestations were more often seen in the operated cats than in the normal ones. A wider dilatation of the right pupil and erection of the hair only on the right side of the body were also observed. On the whole, though tonic manifestations were more pronounced on the decentralized side, clonus was also present and often exaggerated following cerebral ablations. This is in keeping with the findings of Uyematsu and Cobb (1922), Pollock and Davis (1922), Sparks (1927), Davis and Pollock (1928), Riser, Gayral and Pigassou (1946), Wikler and Frank (1948) and others, who showed that, contrary to the opinion held by older authors, the cerebral cortex is not essential for the production of a clonic seizure.

(b)   *Analysis of group responses.* Analyses of the clonic and tonic components of pentylenetetrazol-induced convulsions have been made in experimental animals and in man by numerous investigators (Camp, 1928; Bertha, 1928; Cook and Walter, 1938; Strauss and Landis, 1938; Strauss *et al.*, 1939; Liebert and Weil, 1939; Pollock *et al.*, 1939; Rubin and Wall, 1939; Asuad, 1940; Orloff *et al.*, 1949; Radouco *et al.*, 1952 a, b; and others). On the whole, the pattern of the convulsions described by these authors fits in with those seen by Drake, Seguin and Stavraky (1956), but no conclusive study was made of the incidence of CL and CTC convulsions in relation to doses of pentylenetetrazol and no data pertaining to the duration of the various components of the convulsions or of the convulsibility of control and operated animals were available. When the clonic and CTC convulsions were studied separately (Drake, Seguin and Stavraky, 1956), it became evident that an increased dose of pentylenetetrazol resulted in an increased frequency of CTC convulsions and a corresponding decrease in the percentage of clonic convulsions, a 12 mgm./kgm. dose of pentylenetetrazol giving a CTC-type of convulsion in 100 per cent of the normal animals. A similar alteration of the convulsive pattern with increasing doses of pentylenetetrazol was noted by Liebert and Weil (1939), Pollock *et al.* (1939), Orloff *et al.* (1949) and others. This finding is not limited to pentylenetetrazol: Sparks (1927), Pike *et al.* (1929) observed it with increasing quantities of absinthe; Pollock *et al.* (1939) with camphor and picrotoxin, while Toman, Swinyard, and Goodman (1946), Goodman, Swinyard, and Toman (1946), Barany and Stein-Jensen (1946) described an analogous relationship in convulsions induced by electro-shock. Smith, Mettler, and Culler (1940) also found this to be true in the case of direct electric stimulation of the motor cortex: weak currents produced clonic seizures, while currents of greater intensity yielded essentially tonic convulsions. Yakovlev (1937) observed the same relationship in epileptic patients in whom minor motor seizures were clonic, whereas severe convulsions were predominantly tonic. Not only were the CTC convulsions more frequent both in the normal and the operated animals when increasing doses of pentylenetetrazol were administered, but the operated cats had a greater percentage of CTC convulsions than the normal animals at each dose level. A separate analysis of the CTC and

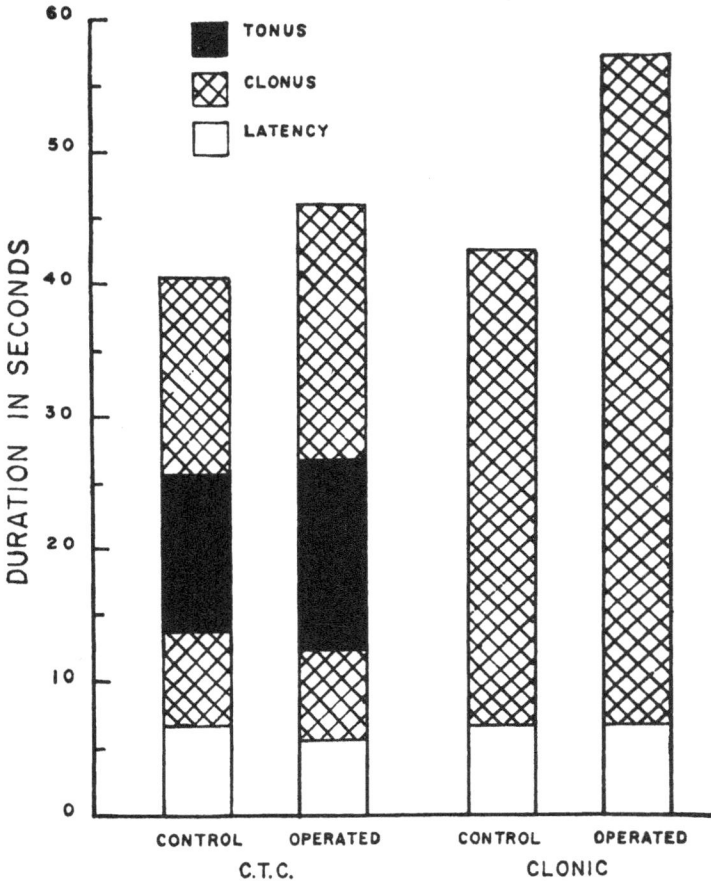

FIG. 13. Diagrammatic representation of responses to pentylene-tetrazol (I.V.) in chronic frontal-lobectomized and semi-decere-brated cats and in control animals. Duration of the latent periods and of various phases of the clonic-tonic-clonic (CTC) and clonic (Cl) convulsions is compared in combined results of 6, 8, and 10 mgm./kgm. doses. (Drake, Seguin and Stavraky, 1956)

clonic convulsions in the two groups of animals gave some interesting results (Fig. 13):

(1) *Clonic-tonic-clonic (CTC) convulsions.* Both the operated and control animals showed a gradual decrease in latency as the dose of pentylenetetrazol was increased. This was most prominent in the controls with the 8 and 10 mgm./kgm. doses, while in the operated cats a

significant reduction of the latent period occurred between 6 and 8 mgm./kgm. In both the intact and operated groups of cats there was wide variation in the duration of the initial clonic phase. However, the operated group had a consistently longer tonic period at each dose level than did the intact animals. The operated cats also had a consistently longer terminal clonus than the intact animals at each dose level. In each group the total duration of the CTC convulsions showed little change with increasing doses of pentylenetetrazol over the range employed, but convulsions were, again, significantly longer in the operated than in the normal animals.

(2) *Clonic convulsions.* The control group of cats showed no significant change in duration of the latent period of the clonic convulsion with increasing doses of pentylenetetrazol. On the other hand, in the operated animals a reduction in the duration of the latent period did take place between the 6 and 10 mgm./kgm. doses. In both the control and operated cats the clonic convulsions became shorter as the dose of pentylenetetrazol was increased, but in the operated group of cats the clonic convulsions lasted longer than in the intact and often appeared to be more severe. The duration of the clonic convulsions seemed, on the whole, to be longer than that of the CTC-type both in the control and in the operated animals.

(3) *Comparison of the convulsibility of the operated and control cats.* From the results accumulated during the study of the types and durations of various phases of convulsion, the median convulsant dose of pentylenetetrazol was calculated. In this calculation the chi-square test for heterogeneity of the frontal-lobectomized and semidecerebrated cats indicated that the convulsibility of the two groups of animals was not significantly different. This and direct observations of the patterns of convulsion justified the treatment of the two groups of operated animals employed in the analysis as a single unit. The clonic and CTC responses were also pooled. Thus calculated, the median convulsant dose of pentylenetetrazol had a maximum likelihood estimate of 7.8 mgm./kgm. for the control cats with values for 95 per cent confidence limits ranging between 7.4 and 8.1 mgm./kgm., whereas for the operated animals the values were 6.8 (6.3–7.2) mgm./kgm. This showed that the general excitability of the operated animals was significantly greater than that of the controls.

ii. *Camphor.* Camphor has been extensively used as a convulsant agent, and for this reason its effects were compared with those of pentylenetetrazol by Drake (1947) and by Drake, Seguin and Stavraky (1956). Camphor monobromate was dissolved in 95 per cent ethanol in accordance with the technique of Wortis, Coombs and Pike (1931). The alcoholic solution of camphor was injurious to the veins and the site of the injection had to be changed on repeated administration of this drug. Also, secondary convulsions were frequently seen after injections of camphor, and because of these factors no detailed analysis of the data was attempted. On the whole, the observations of Wortis, Coombs, and Pike (1931) as to the quantity of camphor required to produce convulsions (3.3 to 6.1 mgm./kgm. for normal cats) were confirmed. The impression was also gained that operated cats were more sensitive to camphor than were intact animals; in two semidecerebrate animals convulsions were produced repeatedly with as little as 2.4 mgm./kgm. of camphor. The convulsions resembled those induced by pentylenetetrazol, but camphor seemed to bring out more prominently the tonic components, and in the operated animals the asymmetry of the convulsion became very marked.

iii. *General remarks pertaining to the action of convulsants.* The foregoing observations show that pentylenetetrazol and camphor, as well as other chemical stimulating agents, exert a greater effect on partially isolated regions of the central nervous system than they do on intact areas. This was demonstrated in electrical recordings of the activity of spinal motor neurones with pentylenetetrazol in the same way as it had been shown with acetylcholine. Thus, there is no doubt that the preceding lesion exerted its effect on these neurones and did not act indirectly through a penultimate sensitization of the muscles. On the other hand, the augmented sensitivity of the corresponding spinal centres after frontal lobectomy or semidecerebration eliminated any possibility that the asymmetrical responses were due to an uneven recovery of the motor neurones from the shock of transection of the cord.

The outstanding finding in the study of the responses of the animals with chronic lesions to convulsant agents was that the median convulsant dose of pentylenetetrazol was reduced by the preceding removal of one frontal lobe or of a complete cerebral hemisphere. This supports unequivocally the contention of Sauerbruch (1913), of Dandy and Elman (1925), and Dandy (1927) that cerebral lesions and ablations intensify chemically induced convulsions.

Statistical studies of the various aspects of the convulsive patterns lend further support to this conclusion. With an increase in the quantity of injected pentylenetetrazol the severity and the frequency of occurrence of the tonic component of the convulsion were increased both in the control and in the operated animals. The tonic phase was more prevalent at each dose level in the latter, beginning earlier in the convulsion, and lasting longer than in the controls. In intact animals similar observations were made previously by Pollock, Finkelman, Sherman, and Steinberg (1939), Goodwin, Kerr and Lawson (1940). Toman, Swinyard and Goodman (1946) also associated the tonic extensor component of the seizure with maximal stimulation of the brain.

The terminal clonus of chemically induced convulsions is often regarded as an interrupted tonic phase (Cobb, 1924; Strauss and Landis, 1938; Strauss, Landis and Hunt, 1939). This phase also was lengthened in duration in the operated cats, and this may be considered as another manifestation of the greater severity of the seizures in the operated group of animals as compared to the controls.

The latent period between the injection of the convulsant drug and the fully developed (CTC) seizure was shorter in the operated animals than in the controls, whereas the mean duration of the convulsions was significantly longer in the operated group. These findings again fit in with the concept that the operated animals were more susceptible to convulsant agents than the controls.

An analysis of photographic records showed that the rolling away from the side of the lesion which occurred in cats with unilateral cerebral ablations during pentylenetetrazol- or camphor-induced convulsions took place during the beginning of the seizure. It was associated with a tonic flexion and withdrawal of the front leg contralaterally to the cerebral ablation and occurred concurrently with clonic extensor thrusts of the limbs on the side of the operation. These extensor thrusts seemed to push the animal away from the side of the operation, the cat falling and continuing to roll to the side which lost its support due to a tonic flexion of the extremities. Occasionally, in the absence of any convulsive manifestations on the other side of the body, very small doses of camphor or pentylenetetrazol caused unilateral extensor thrusts in the front limb opposite the cerebral ablation. In these instances, the animals fell to the operated side. Thus the direction of the rolling and falling of the animals with cerebral ablations, which was previously noted by Pike and Elsberg (1925) and by Dandy and Elman (1925), can be attributed to the exaggerated responses of the denervated side to convulsant agents. The circling of the operated animals, which was occasionally seen either after injection of very small quantities of the drugs or preceding or following a convulsion, may also be explained in this way.

It should be stressed that in the chronic experiments the responses of the denervated side following the injection of various stimulating agents resembled co-ordinated movements induced by cortical or reflex stimulation. When the dose of the drug was minimal, the movements were unilateral and often of an extensor type. When the quantity of pentylene-tetrazol or camphor was excessive, the movements were of a tonic flexor nature with automatic movements which often resembled a grasp reflex appearing in the after-effect. In both instances, however, the convulsant drug evoked at the height of its effect something like a "reflected image" of the response which normally had originated in the motor cortex.

It is of interest to note that the asymmetry during pentylenetetrazol- or camphor-induced convulsions in the operated animals was always more pronounced in the fore limbs than in the hind limbs. This fits in with the findings of Lassek (1946), who demonstrated that in the cat the majority of the pyramidal fibres terminate in the cervical segments of the cord. An alternative explanation for the greater asymmetry of responses in the fore limbs may possibly be the fact that the labyrinthine and neck reflexes exert a synergistic effect on the upper extremity but oppose each other in the lower limb (Magnus and de Kleijn, 1912). Finally it may be mentioned that the increased sensitivity of frontal-lobectomized cats was not limited to chemical stimulating agents, but was demonstrated by McKenzie, Seguin and Stavraky (1960) with electric stimulation of the brain. These investigators found that a significant mean decrease in the electro-shock threshold occurred one to six months after the operation, while subthreshold currents which caused no generalized convulsions produced extensor tonus of the limbs contralaterally to the removed frontal lobe. Electronarcosis could be also induced much more easily in the operated animals than in the controls, was deeper on the decentralized side of the body, and produced rolling of the animals when the current was increased or reduced.

4. NEURONAL ISOLATION OF THE CEREBRAL CORTEX AND CONVULSIBILITY IN ANIMALS AND MAN

A. *Section of the corpus callosum*

The over-all increase in the convulsibility of the frontal-lobectomized and semidecerebrated animals strongly suggested the possibility that not only the spinal motor neurones but the higher levels of the central nervous system as well were rendered supersensitive by the cerebral ablations. A confirmation of this impression was provided by Teasdall (1950) and by Teasdall and Stavraky (1950) in an investigation carried out on twenty cats in which the corpus callosum was aseptically divided. After recovery from the operation, the animals were periodically convulsed with acetylcholine and pentylenetetrazol over several months.

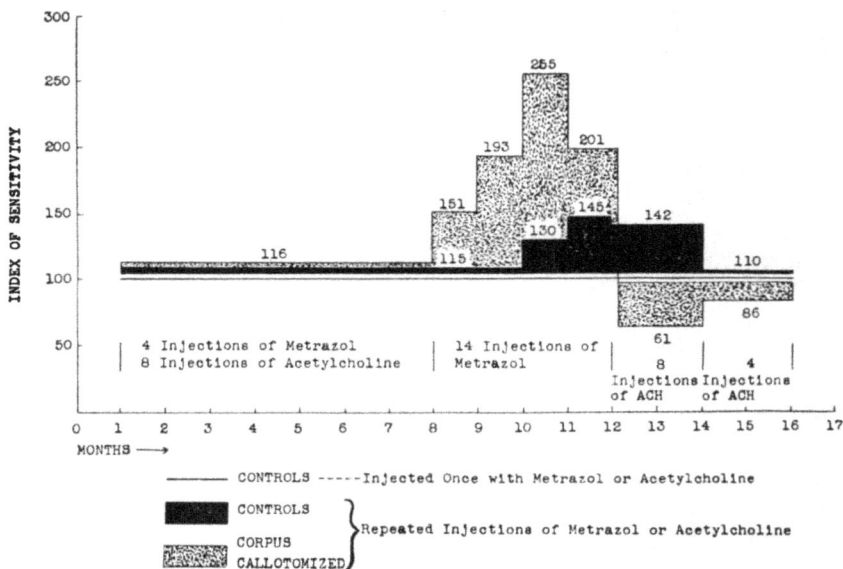

FIG. 14. Variations in sensitivity of eight control and ten corpus-callotomized cats during a prolonged period of injections of pentylenetetrazol (metrazol) and of acetylcholine. The sensitivity of the control cats injected once with pentylenetetrazol or acetylcholine is taken as 100. The index of sensitivity of the corpus-callotomized cats, and control animals injected repeatedly simultaneously with them, is expressed relative to the index calculated for the controls injected once (see text). (Teasdall, 1950)

The sensitivity of the corpus-callotomized cats, as judged by the percentage of animals convulsing per unit of analeptic agent, was compared to the sensitivity of two control groups of cats. One group of controls received only one injection of either pentylenetetrazol or acetylcholine at each dose level, while the other group received repeated injections of the convulsant agents in the same quantities and at the same time intervals as the corpus-callotomized animals. The relative sensitivities of the three groups of animals are represented in Fig. 14 by means of an arbitrary index of sensitivity in which the convulsibility of the control cats injected once is compared with the convulsibility of the two groups of cats which were injected repeatedly. The intervals differed at particular stages of the experiment, but at no time were the animals injected more frequently than once a week. No striking difference between the control and corpus-callotomized animals could be detected until the eighth or ninth month following the operation. However, after this prolonged recovery period, when the animals were injected at weekly intervals with various quantities of pentylenetetrazol, the convulsibility of the corpus-callotomized cats became progressively greater. This increase in sensitivity became maximal after six or eight injections, at which time the threshold of the animals to pentylenetetrazol was diminished by twenty to fifty per cent. The convulsive pattern of the corpus-callotomized cats was also altered in a characteristic way: the duration and the severity of the clonic phases of the convulsion were markedly increased as compared with that of the tonic period. This resulted in a prolongation of the total duration of convulsions and a status epilepticus, which lasted one to twenty-seven hours after the injection of 6 and 8 mgm./kgm. of pentylenetetrazol, was frequently seen. In fact, so severe were the seizures that four corpus-callotomized animals in the first tested group died in status epilepticus. Another feature of the convulsions was the extreme prostration that occurred following the convulsions, the animals lying on their sides unable to raise themselves for prolonged periods of time. The duration of the post-convulsive depression, as determined by the return of the placing reactions, was found to be longer in the corpus-callotomized cats than in the control animals. The absence of placing reactions following experimental convulsions had been noted in intact animals by Ward and Clark (1938) and by Dille and Hazelton (1939), and is attributed to the depression of the cortical activity.

After four months of weekly injections of pentylenetetrazol, the corpus-callotomized animals became partially refractory to small doses of both pentylenetetrazol and acetylcholine. On the other hand, the convulsibility of intact control cats was not adversely affected by similar repeated injections of pentylenetetrazol, and their sensitivity to pentylenetetrazol increased somewhat instead of being depressed. A similar observation for intact animals was made by Sacks and Glaser (1941), who found that rats became more sensitive to pentylenetetrazol following weekly injections.

After a rest of one year, the sensitivity of the corpus-callotomized cats to convulsant drugs was found to be greatly exaggerated again. It was an orientational experiment, but the results were interpreted by Teasdall (1950) and Teasdall and Stavraky (1950) to indicate that a greater sensitivity of the partially denervated cerebral cortex to chemical stimulating agents takes place after sectioning of the commissural fibres joining the two cerebral hemispheres. These conclusions were later substantiated statistically by Seguin, Fretz and Stavraky (1957) and by Seguin, Fretz, Manax and Stavraky (1961) in Sprague-Dawley rats.

The gradual onset of the sensitization following section of the corpus callosum should be emphasized. This requires not only time for its onset following the operation, but it develops progressively with repeated injections and seems to run in a cycle which is different from that observed in control animals. In this connection it is interesting that Hoefer and Pool (1943) found the convulsibility of cats and monkeys to be reduced for as long as two weeks after the section of the corpus callosum. However, the use of anaesthesia may have influenced the results in these experiments.

The opinion is often expressed that no detectable signs are produced by surgical section of the corpus callosum in man (Dandy, 1930; Armitage and Meagher, 1933; Akelaitis, Risteen, Herren and van Wagenen, 1942; Smith and Akelaitis, 1942; and others). On the other hand, it is established experimentally that, although it is not directly concerned with motor activity, the corpus callosum is associated with movements initiated from the cerebral hemispheres, and that its section results in a deficit in interhemispheric transfer of somesthetic discrimination and of learning (Stamm and Sperry, 1957 and Meyers and Sperry, 1958). In spite of the fact that generalized epileptiform convulsions can take place on unilateral stimulation after corpus callotomy (Karplus, 1914; Obrador, 1942; see Bremer *et al.*, 1956), as early as 1890 Mott and Schaefer described movements of the extremities induced in monkeys by electrical stimulation of this structure. Kennard and Watts (1934) found that section of the corpus callosum in this species caused inertia and slowness of movement which persisted for several days. Seletzky and Gilula (1928) localized these effects to the anterior part of the corpus callosum in rabbits and dogs. Slowness of movement and dyspraxia were also described on section or involvement by tumours of the anterior part of this structure in human subjects (Liepmann and Maas, 1908; Wilson, 1908; Ironside and Guttmacher, 1929). The electrical activity of one cerebral hemisphere in response to local application of convulsant agents and electric stimulation of the contralateral cortical areas was studied by Gozzano (1936), Moruzzi (1939, 1952), Curtis (1940 a, b), McCulloch and Garol (1941), Hoefer and Pool (1943), Garol (1942), Kopeloff *et al.* (1950), Chang (1953), Peacock (1957), Grafstein (1957, 1959) and others; and the existence of homotopical and heterotopical functional connections between the two cerebral hemispheres by way of the corpus callosum and anterior commissure was confirmed. A morphological basis for these observations was provided by Pines and Maiman (1939), who found evidence of retrograde degeneration in areas 4, 6, 8 and 12 of the frontal cortex following section of the corpus callosum in dogs and in patients with tumours involving this structure. On the basis of this latter investigation as well as of his own observations, Erickson (1940) suggested that some fibres of the corpus callosum may be collaterals of the pyramidal tracts.

Assuming that the spread of the epileptic discharge takes place from one cerebral hemisphere to the other by way of the corpus callosum, van Wagenen and Herren (1940) sectioned this commissural tract in ten cases of generalized epilepsy in an attempt to localize the seizures to one cerebral hemisphere. The patients were studied for up to five months after the operation, and conversion of the attacks to unilateral convulsions was reported in eight of them, while in the remaining two patients the generalized seizures persisted. On the whole, van Wagenen and Herren (1940) thought that the seizures were less severe after the section of the corpus callosum. It must be remembered, however, that they based their opinion on a study conducted during a five-month interval after section of the corpus callosum, while the increased sensitivity of the experimental animals in the investigations of Teasdall (1950) and Teasdall and Stavraky (1950) did not develop until eight or nine months after the operation. Furthermore, if barbiturates were administered to these patients during the post-operative period, this fact could readily account for the diminished severity of the convulsions. A more profound depression by barbiturates of the cortical neurones after their isolation by section of the corpus callosum would be expected from the experimental findings that the depressant effects of the sodium pentobarbital and phenobarbital are enhanced by denervation (Drake and Stavraky, 1948 b; Seguin and Stavraky, 1952, 1954; Seguin, 1956).

Finally, it must be kept in mind that the patients in whom the corpus callosum had been sectioned still had recurrent attacks of epilepsy; it is conceivable that cortical neurones which were sensitized by section of the corpus callosum also became more readily depressed by repeated epileptic discharges and a self-limitation of the severity of the convulsions took place. An experimental basis for this contention may be found in the fact that during the post-convulsive period experimental animals become partially refractory to convulsant procedures (Elsberg and Stookey, 1923; Coombs, 1932; Fender, 1937); Teasdall (1950) and Teasdall and Stavraky (1950) found the post-convulsive depression to be greatly exaggerated and prolonged in corpus-callotomized cats.

On the other hand, it may be pointed out that the most striking feature of Marchiafava's disease of primary degeneration of the corpus callosum in human subjects is the occurrence of spontaneous motor seizures (Bignami and Nazari, 1915; King and Meehan, 1936; Bohrod, 1942;

Erickson, 1940). Convulsions were observed not only in clinical cases of primary degeneration of the corpus callosum but also when tumours of the corpus callosum were found (Beling and Martland, 1919; Marchand and Schiff, 1925). Thus, there is good reason to believe that destruction of the corpus callosum results in an ultimate lowering of the convulsive threshold in human beings as well as in experimental animals. Of interest also is the finding that the severity and the duration of the clonic phases of the convulsions increased after section of the corpus callosum. Although it was shown repeatedly that clonus may take place in the absence of the motor cortex, the importance of this region of the brain in the production of the clonic phases of epileptiform seizures is generally recognized. The evidence for this was fully reviewed by Moruzzi (1946). If this view is accepted, the prominence of the clonic manifestations after section of the corpus callosum may be interpreted as due to a sensitization of the cerebral cortical regions concerned with the production of clonus.

B.   *Isolation of the motor cortex*

The exaggerated irritability of the motor cortex after its partial isolation manifests itself not only in an increased effectiveness of chemical stimulating agents, but also in a greater sensitivity to electrical stimulation. This was shown by Bard (1933) in cats with a decortication of one cerebral hemisphere except for the gyrus proreus and the anterior and posterior sigmoid gyri. When explored electrically several weeks later, the isolated motor cortex had a lower threshold than that on the intact side, and the movements elicited from the isolated motor area were more vigorous and prolonged. Obrador (1947), in an analogous experiment on dogs isolated an area of the motor cortex of about two centimetres in diameter by a circular subpial incision so as to preserve the main blood supply. Adhesions were prevented by placing a thin piece of gutta percha bilaterally over the exposed areas. At the end of two months the cerebral cortex was exposed once again and its excitability tested by means of electrical stimulation. The isolated cortex had a lower threshold compared to the opposite side; with higher than threshold voltages, its responses became epileptiform in nature. Local application of weak solutions of strychnine to the two sides resulted in a clonic discharge which was limited to the limbs contralateral to the isolated area. The clonus was sustained and had a regular rhythm of about sixty to seventy contractions and relaxations per minute. Removal of the isolated cortical area abruptly stopped the clonus. These observations were expanded by Echlin and McDonald (1954), who isolated blocks of cerebral cortex "neuronally" by a subpial division and undercutting in monkeys and in a patient suffering from schizophrenia. Three to nine months later, the cerebral cortex was exposed under light ether and local anaesthesia, and the activity of the isolated region was studied by means of electrocorticograms. The partially isolated cortex had a resting discharge of random spikes or bursts of rhythmic high-voltage spikes and slow waves similar to those seen in a focal epileptogenic lesion. The application of acetylcholine caused rhythmic high-voltage spike and wave discharges which again occurred selectively in the isolated area; with 0.5–1.0 per cent acetylcholine the isolated area was affected before and more profoundly than the surrounding cortex, but some spike discharges occurred from elsewhere. In two animals application of 1 per cent acetylcholine to a partially isolated area caused focal motor seizures in the opposite extremities which lasted as long as five days. Preceding eserinization potentiated and prolonged the effects of acetylcholine, whereas the application of the same or stronger concentrations of acetylcholine to acutely isolated areas in control experiments on intact monkeys had no effect. Intravenous injections of discriminative doses of pentylenetetrazol caused selective activation of the chronically isolated cortex, and prolonged spike and wave discharges were observed to occur only in the isolated zone. Similarly, electrical stimulation of the isolated region resulted in after-discharges which were greatly prolonged compared to those of the homologous area of the opposite hemisphere.

In the human subject an area of the left frontal lobe was partially isolated by subpial section along three sides and undercutting. Ten months later, under light pentothal anaesthesia, both frontal lobes were exposed in preparation for lobotomy, and electrocorticograms were taken. Irrigation of the frontal lobes with acetylcholine solution resulted in a high-voltage, rhythmic spike discharge on the isolated side, but acetylcholine had no effect on the opposite frontal cortex. The electrical discharges were so alarming on the side of the preceding isolation that the patient was put under deep pentothal anaesthesia.

These studies in the opinion of Echlin and McDonald (1954) provided "direct evidence that the denervated or partially denervated cerebral cortex is supersensitive to certain forms of stimuli." A further expansion of these observations and a detailed electroencephalic analysis of the behaviour of neuronally isolated cerebral cortex in monkey and man were recently provided by Echlin (1959) in a publication which related "isolation phenomena" to the mechanism of focal epilepsy.

Prolonged after-discharges and persistent "spontaneous" activity of neuronally isolated slabs of cerebral cortex have been reported by

Burns (1951, 1954, 1958), while Halpern and Sharpless (1959) studied the onset of the increased duration of after-discharge and the lowering of the threshold elicited by direct electrical stimulation in the undercut slabs of suprasylvian cortex of cats with chronically implanted electrodes. Supersensitivity of such isolated cortical slabs according to the latter investigators developed within three weeks after the operation, the homologous area of the opposite hemisphere also showing some increase in excitability compared to adjacent portions of intact cortex.

5.   SENSITIVITY OF PARTIALLY ISOLATED NEURONES TO ANTICONVULSANT AGENTS

A.   *Effects of sodium pentobarbital*

i.   *Intra-arterial injections in spinal cats.* It was established in several investigations that partially isolated neurones of the central nervous system become more sensitive to depressant drugs as well as to stimulating agents. In the case of barbiturates, this was observed first by Stavraky (1947) during the study of the effects of adrenaline in animals with chronic lesions of the central nervous system. Drake and Stavraky (1948 b) found the same to be true after chronic deafferentation, and Seguin and Stavraky (1952, 1954, 1957) and Seguin (1956) investigated this apparently selective action of barbiturates on isolated neurones in greater detail. The susceptibility of spinal neurones to barbiturates was tested on responses of the quadriceps in high-spinal cats in which the spinal cord had previously been semisected at the T12 level. The excitability of the two sides was tested by means of intra-aortal injections of acetylcholine. Small quantities of sodium pentobarbital were occasionally seen to have a slight excitatory action on spinal neurones; larger quantities of this agent exerted a depressant effect. The minimal effective dose of sodium pentobarbital varied in different experiments, but was always less for the previously semisected side compared with the control side. Thus, when discriminative doses of sodium pentobarbital were injected, a selective depression of the chronically denervated spinal neurones occurred which often resulted in a reversal of the effect of acetylcholine on the two sides of the spinal cord similar to that found by Drake and Stavraky (1948 b) with deafferentation.

ii.   *Intravenous injections in cats with brain lesions.* The exaggerated action of barbiturates on partially isolated regions of the central nervous system was also investigated in chronic animals (Seguin, 1956; Seguin and Stavraky, 1957). Sodium pentobarbital was injected intravenously into left-frontal-lobectomized or semidecerebrated cats and into controls. In the control animals injections of 2 to 8 mgm./kgm. of sodium pentobarbital resulted in an unsteady gait, the cats falling and staggering with equal frequency to either side; with the higher doses they were often unable to stand or walk for periods of time ranging from two to forty-six minutes.

In spite of a considerable difference in the susceptibility of individual animals, the injection of even the smallest tested quantities (1.2 mgm./kgm.) of sodium pentobarbital caused the operated cats invariably to lean, stagger, or fall away from the side of the operation. In the case of left-sided lesions, the animals dragged their right forepaws when walking, and stepped on their dorsa, while the right hind paws slid out from under them. When larger quantities of sodium pentobarbital (2.0–8.0 mgm./kgm.) were injected, a greater weakness and inco-ordination developed on the right side, and the animals were unable to stand unassisted for up to 120 minutes. This inability of a limb to support the weight of the animal was taken as a criterion of depression. The "mean depression time," determined individually for each of the four extremities, was the same on both sides in the control group, and was shorter than that on either side in the operated group. Furthermore, while in the control animals the limbs on the two sides were depressed symmetrically, in the operated cats the right extremities were depressed for much longer periods of time than the left. These differences were statistically significant.

The selective depression of the limbs contralaterally to the cerebral ablations is shown in Figs. 15 and 16. The photographs were taken during the period of recovery from the bilateral depression caused by sodium pentobarbital. In order to demonstrate the depression most effectively, the cats were partially supported so that their weight rested on the tested extremities. During the recovery the inco-ordination and weakness of the corresponding limbs was also clearly visible when the animals attempted to walk.

a) *ED_{50} of control and operated cats.* Because of the apparently greater susceptibility of cats with cerebral ablations to the depressant action of sodium pentobarbital, the median effective doses (ED_{50}) of this drug were determined by Seguin and Stavraky (1957) for control cats and for cats in which the left frontal lobe was removed twenty-four to thirty-six months previously. The "standing test," defined as the inability of the animals to stand, was taken as a measure of the effective-

Fig. 15. Effects of 5 mgm./kgm. sodium pentobarbital (I.V.) in a cat twenty-two months after the removal of the left frontal lobe including the motor cortex. Before the injection, the animal walked normally, the only after-effect of the operation being the absence of the "placing reaction" in the right limbs. When raised by the hind limbs the cat supported its weight equally on both fore paws (A). After the injection the cat was unable to walk for twelve minutes. When during this time the pelvis was raised the cat supported its weight only on the left fore paw, the right extremities remaining limp (B). During the following twenty-five minutes, the animal walked, leaning, staggering and falling to the right; marked weakness and inco-ordination of the right limbs was evident.
(Seguin and Stavraky, 1957)

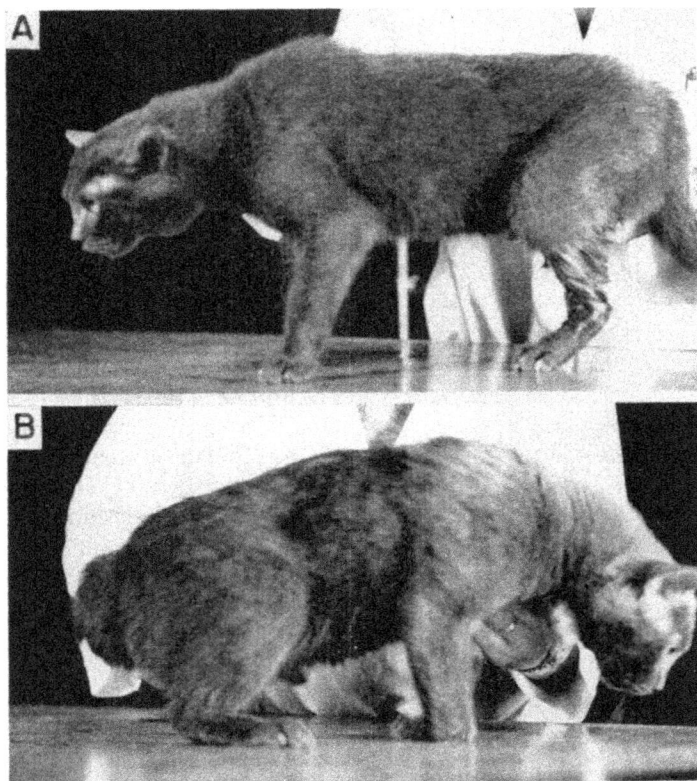

FIG. 16. Effects of 5 mgm./kgm. of sodium pentobarbital (I.V.) in a cat twenty-one months after the removal of the complete left cerebral hemisphere. Before the injection the animal walked almost normally but circled to the left (as in Fig. 5 (5)). Eight seconds after the injection the cat fell to the right and was unable to stand or walk for twenty minutes, showing signs of weakness and inco-ordination in the right limbs for half an hour. When supported from the right side it could stand on the left extremities (*A*); when supported from the left side the right limbs collapsed under the weight of the body (B). (Seguin and Stavraky, 1957)

ness of sodium pentobarbital. This test was chosen because it offered the most definite end-point of any of the effects of sodium pento-barbital common to both the operated and control cats. The ED_{50} for the control cats was 3.7 (2.5–5.6) while for the operated animals it was 1.7 (1.1–2.5), the values in parentheses representing 95 per cent confidence limits. Thus sodium pentobarbital was found to depress the operated cats 2.2 times as much as the controls.

B. *Effects of other anticonvulsants*

Sodium phenobarbital affected the operated cats in a similar manner; however, larger doses of the drug were needed than those of sodium pentobarbital to produce a comparable degree of inco-ordination and the effects came on more gradually, lasted longer, and were less clearly defined.*

It is interesting that Morita (1915) found decorticate rabbits to be more affected by urethane and chloralose than intact ones. Mettler and Culler (1934) noted in a completely decorticate dog a greater susceptibility to morphine and anaesthetics after recovery from the operation than in a normal animal, and Wikler (1950) found the same to be true in the case of barbiturates in chronic experiments on decorticate cats. Dasgupta, Mukherjee and Werner (1954) found that chlorpromazine and pentothal depressed the "rage reaction" in decorticate cats more markedly than in intact animals, this effect becoming apparent as early as forty-six hours after the removal of the cortex. An interesting observation was also made by Dmitriev (1953) on reptiles and amphibia in which a transverse section of the spinal cord led to a temporary paralysis of the hind limbs; when motor activity in the affected extremities was restored, chloral hydrate or urethane in appropriate quantities selectively affected the movements of the lower parts of the body. This indicated that the isolated portion of the spinal cord acquired a greater sensitivity to depressant agents than the rest of the central nervous system.

C. *Effects of barbiturates on pentylenetetrazol-induced convulsions*

The way in which sodium pentobarbital influences pentylenetetrazol-induced convulsions was studied by Seguin and Stavraky (1957) in chronic frontal-lobectomized and semidecerebrated cats. Pentylenetetrazol was injected once a week in combination with increasing quantities (1, 2, 4, and 6 mgm./kgm.) of sodium pentobarbital given intramuscularly one hour before the pentylenetetrazol. It was found that a dissociation of the convulsive patterns often took place. The hind limbs developed a clonus while the front limbs went through a tonic phase. In cats with left-sided ablations, the initial tonus in the right limbs was often absent, a symmetrical clonus marking the beginning of the convulsion. In some animals this was the only effect seen, but in others a uni-

---

*In a recent study on rats with various brain lesions, Adler, using the duration of sleeping time and the convulsion time as measures of sensitivity to hexobarbital and to pentylenetetrazol, fully substantiated the findings presented in this chapter. (Adler, M. W. Changes in sensitivity to hexobarbital and pentylenetetrazol after brain lesions in rats. The Pharmacologist 2: 76)

lateral clonus on the left side was observed, with a complete suppression or shortening of the convulsion on the right side of the body. These symptoms indicated a reversal of the usual trend. In some animals the rolling also showed a tendency towards reversal: the cat began to roll to the right side, but later in the convulsion changed its direction and rolled to the left. The number of convulsions induced by pentylenetetrazol decreased progressively in both the operated and control groups as the dose of sodium pentobarbital was increased. However, at 1 and 2 mgm./kgm. doses the total number of convulsing animals was greater in the operated group, whereas with 4 and 6 mgm./kgm. of sodium pentobarbital this trend was reversed. Particularly striking was the decrease in the number of CTC convulsions in the operated cats under the influence of sodium pentobarbital. When more than 2 mgm./kgm. were given, all the convulsions in the operated group became clonic; the same result in the control group was reached only at the 6 mgm./kgm. dose level.

The latent period between the injections of pentylenetetrazol and the convulsions was not significantly altered in the controls by sodium pentobarbital. On the other hand, in the operated cats it became progressively longer with increasing doses of sodium pentobarbital. Similarly, the duration of the convulsions was markedly lengthened in the operated cats when pentylenetetrazol was preceded by sodium pentobarbital, whereas in the control group this prolongation of the seizures was not significant. The increase in the duration of the convulsions was apparently due to the fact that a large number of convulsions were changed from a relatively short but intense CTC type to a more prolonged but less severe clonic seizure. In one of the earliest studies of the effects of barbiturates on chemically induced convulsions, Sparks (1927) showed that sodium phenobarbital and amytal could change a thujone-induced seizure from a clonic-tonic-clonic type to a purely clonic one. A depression by sodium phenobarbital and some other barbiturates of the tonic component of the pentylenetetrazol-induced convulsions was also seen by Goodman, Grewal, Brown and Swinyard (1953), while the relatively longer duration of the clonic convulsions induced by pentylenetetrazol, as compared to the CTC type, was demonstrated by Drake, Seguin and Stavraky (1956). Once the change-over from the CTC to the clonic type was completed, the gradual reduction in the duration of convulsions with larger doses of sodium pentobarbital became apparent; it depended on a progressive diminution of the severity of the clonic convulsions. Such an effect of barbiturates on the clonic phases of pentylenetetrazol-induced convulsions in intact animals was demonstrated by Frommel and Radouco-Thomas (1953) and others and would account for the findings in the operated group as well.

6.   CONCLUSIONS AND CORRELATION OF EXPERIMENTAL RESULTS
WITH SOME CLINICAL DERANGEMENTS OF THE CENTRAL
NERVOUS SYSTEM

In summary, it may be stated that after partial cerebral ablations, there develops in the remaining portions of the central nervous system an increased sensitivity both to chemical stimulating agents and to depressant drugs. This is also true for electrical stimulation of the brain as judged by the increased convulsibility of frontal-lobectomized animals and by the responsiveness of the isolated motor cortex. The "supersensitivity" which develops following partial isolation is more pronounced in regions directly connected with parts of the brain subsequently removed. However, the supersensitivity appears in areas sufficiently far removed from the site of the operation to exclude any possibility of scarification or glial overgrowth being involved in the process, occurring, for example, in the anterior horn cells after excision of the motor cortex. It comes on gradually, in some instances reaching its peak weeks and even months after the operation, and once established is expressed in a shortened latency, lowered threshold, often greater and more sustained responses, and longer after-effects. As pointed out by Cannon and Rosenblueth (1949, chapter II, pp. 11–21) the "reactivity" or degree of response in the peripheral structures is not directly increased by denervation. The apparent exaggeration of different responses brought about by the stimulation of the central nervous system may be of a complex nature, involving the activation of greater numbers of neurones, or summation of prolonged after-discharges in reverberating circuits, or both, rather than more powerful responses of the individual units themselves.

A.   *Sensitization by denervation and epilepsy*

Considering convulsibility in the light of clinical disorders, it is known that epileptic patients are more susceptible to chemical convulsant agents than normal human beings: this is particularly well substantiated in relation to pentylenetetrazol (Kaufman, Marshall, and Walker, 1947; Walker, 1949; Kalinowsky and Kennedy, 1941; Toman and Goodman, 1947; von Meduna, 1935; Kalinowsky, 1947; Sal y Rosas, 1943, 1945; Goldstein and Weinberg, 1940; Ziskind and Bercel, 1947). Similarly, Pacella, Kopeloff, and Kopeloff (1947) and Chusid, Kopeloff, and

Kopeloff (1953) found that smaller quantities of pentylenetetrazol produced convulsions in monkeys which had epileptogenic lesions induced by the application of alumina cream to the cerebral cortex. Cure, Rasmussen and Jasper (1948), using a similar procedure, showed that subconvulsant doses of pentylenetetrazol activated first the epileptogenic lesion, and that electrically recorded epileptic discharges spread from this cortical focus to other parts of the brain. Similar observations were made in epileptic patients in whom small quantities of pentylenetetrazol were slowly injected into a vein.

On the other hand, the studies of Penfield (Penfield and Jasper, 1954, pp. 241–6), Walker (1949, p. 61) and others have established the fact that in patients with focal epilepsy the cerebral cortex in the vicinity of the lesion is more sensitive to electrical stimulation than it is in normal people. Furthermore, it was found that the area which gives rise to an abnormal discharge and which is most susceptible to the induction of an epileptic seizure is not situated in the most severely damaged part of the brain, but is usually located at its periphery around the scar or new growth. A similar exaggeration of sensitivity in the vicinity of lesions in chronic patients with spinal cord injuries was noted by Scarff and Pool (1946). They found during surgical exploration that below the site of transection in the cord there was a "transition" zone in which minimal or mechanical or electrical stimuli initiated mass contractions which were greater than those obtained when the same stimuli were applied to more distal parts of the spinal cord. Partially isolated areas of the grey matter of the brain also showed paroxysmal hypersynchronous discharges in electrical recordings of their spontaneous activity. Such discharges were comparable to those seen in epilepsy by Echlin *et al.* (1952) and were noted also in neurones situated over a brain tumour. The latter investigators draw attention, however, to the possibility that partial isolation of the cerebral cortex may result, after a transient depression, in some increase of sensitivity to electrical stimulation and to pentylenetetrazol within four to six hours. As mentioned previously (p. 40), Burns (1951, 1954, 1958) and Halpern and Sharpless (1959) found that application of electrical stimuli to the isolated cortex in the cat could result in a prolonged series of discharges of a discrete nature. Burns concluded that the repetitive response is the consequence of self-reexcitation in closed chains of neurones, and his findings may have an important bearing on the mechanism involved in an epileptiform discharge.

It is possible that a pathological process of a focal type, or even a widespread derangement of the brain, may result in a partial isolation of some nerve cells. This is most likely to occur in the vicinity of the lesion as the result of degeneration of small, short-axon neurones. The surviving, large nerve cells would be rendered supersensitive by such a process to chemical agents or to nerve impulses which reached them by way of the remaining connections. Similarly, destruction of long-path neurones, which results in a degeneration of association, commissural, or descending paths, will sensitize more remote regions of the central nervous system. The end-result of isolating a focus or of a widespread area of the brain in this way is a lowering of the threshold of convulsibility, and when this threshold is exceeded, a seizure results which characteristically begins in the sensitized part of the brain. This possibility was considered by Cannon (1939), Cannon and Rosenblueth (1949). Obrador (1947), and Echlin *et al.* (1952, 1954, 1959) and is supported by the present investigations. A detailed consideration of the factors which may precipitate a convulsion is outside the scope of this discussion. Toman and Goodman (1948) list among them changes in pH, in partial pressures of oxygen and carbon dioxide, in total osmotic pressure and electrolyte composition of the fluid environment of the brain cells, as well as changes in body temperature. Gellhorn (1953) called attention to the role of the hypothalamus and re-emphasized the fact that afferent impulses may precipitate or inhibit convulsions. Penfield and Humphreys (1940) and Penfield and Erickson (1941) suggested that a periodic impairment of blood flow occurs in or about the epileptogenous zone. On the other hand, Pope *et al.* (1947) noted an increased cholinesterase activity in the epileptogenic cortex. Tower and McEachern (1949 a, b) showed that acetylcholine was present in the cerebro-spinal fluid of 77 per cent of epileptic patients, but in only 15 per cent of patients suffering from other neurological conditions; and Tower and Elliott (1952) found increased amounts of acetylcholine in epileptogenic areas of the brain. These findings are pertinent in view of the fact that acetylcholine can cause epileptiform convulsions (Miller, Stavraky and Woonton, 1940; Merritt and Brenner, 1941; Stavraky, 1943; and others) and, as shown by Forster and Madow (1950), can sensitize the brain to other epileptogenic agents such as pentylenetetrazol.

While widespread gliosis of the brain in the experimental ablations carried out by Dandy and Elman (1925), Stavraky (1943, 1947), Drake, Seguin and Stavraky (1956) and McKenzie, Seguin and Stavraky (1960) is out of the question in view of the careful asepsis and modern neurosurgical technique employed in their operative procedures, the question remains: how does the removal of a scar or of a new growth lead to a cure in clinical cases of Jacksonian epilepsy? Probably this is achieved partly by the removal of the mechanical irritation which aggravates the condition and partly by removal of the low-threshold zone which lies in the vicinity of the lesion. But even at best, such an excision denervates a new set of neurones somewhere in the central nervous system, and one may expect that the convulsive threshold of the patient will be reduced as the sensitization of the newly denervated area develops. An indication that this actually does occur may be found in recent observations of Lomas *et al.* (1955), Schlichter *et al.* (1956), Liddell and Retterstöl (1957) and others, who found that chlorpromazine brings about epileptic fits almost exclusively in patients who had leucotomies. This is possibly why neurosurgeons, who have had the widest experience with removal of epileptogenic foci, insist that their patients take barbiturates for prolonged periods after an operation and make their patients avoid rigidly any and all aggravating conditions. A progressive sensitization of widespread areas of the brain would also account for the incomplete success in attempted localization of epileptic discharges by means of section of the corpus callosum (operation of van Wagenen and Herren, 1940).

The greater effect of barbiturates and other depressant agents on denervated neurones deserves special comment. Once established, this fact makes it possible to account for the beneficial effect that these drugs exert on epileptic patients. If a barbiturate should affect all the nerve cells equally, one would expect that the most irritable area of the brain, such as an epileptogenic zone, would be the last one to be depressed by the drug. Yet barbiturates, in doses which cause no derangement of normal motor manifestations in the patients, often prevent the occurrence of epileptic attacks. Selective depression of sensitized neurones in the epileptogenic zone of the brain by barbiturates may well account for this fact, and one is tempted to suggest that many other pharmacological agents may exert their effects in various derangements of the nervous system through a preferential action on neurones rendered supersensitive by disease. An example of this possibly would be the benefit derived by patients with a Parkinsonian or post-encephalitic syndrome from atropine, belladonna, hyoscine, and other drugs which in the usual therapeutic doses do not affect a normal central nervous system to any comparable extent.

B. *Convulsive Therapy*

The value of artificially induced convulsions, brought about either by pentylenetetrazol (von Meduna, 1935) or by electro-shock (Cecil, 1944) is well known in the treatment of mental disorders. However, the way in which this widely adopted clinical procedure exerts its beneficial effects is poorly understood at the present time. Less widely recognized is the curative possibility of convulsive therapy in epilepsy. Yet, this form of treatment was employed by Kalinowsky (1947) with success in the control of some forms of predictable seizures: in two patients with regularly occurring "menstrual epilepsy" Kalinowsky could prevent the occurrence of spontaneous attacks by inducing at the appropriate time one or two electro-shock convulsions.

Elsberg and Stookey (1923), Coombs (1932) and Fender (1937) showed that during the post-convulsive period of depression animals were rendered partially refractory to further convulsant procedures. Fischer (1932), Gibbs, Davis and Lennox (1935), Hall (1938), van Harreveld and Stamm (1954) and others observed a transient quiescence, and other changes of electrical activity in the cerebral cortex following seizures both in animals and in epileptic patients. Thus, the conclusion was drawn that this post-convulsive depression was of a cerebral cortical origin, and the recent studies of the "spreading depression" which occurs following repeated electrical stimulation of the exposed brain may be interpreted as favouring this view. Further investigation by de Morsier, Georgi and Rutishauer (1938), Liebert and Weil (1939), Sacks and Glaser (1941) and others revealed that not only immediately following the convulsion is the threshold raised, but that when about ten injections of pentylenetetrazol are repeated two or three times a week, experimental animals develop a prolonged resistance to the drug, and the quantity of pentylenetetrazol has to be increased in order to produce a convulsion.

This "resistance" was found by Stavraky (1943, 1947) to develop not only when true analeptic drugs were employed, but also when convulsions were induced by acetylcholine and adrenaline in frontal-lobectomized and semidecerebrated cats. In the case of acetylcholine and adrenaline the organism is well equipped to destroy these substances rapidly, and it seems more reasonable to assume that the long-lasting depression was caused by a change in the sensitivity of the nerve cells brought about by a rapid succession of excessive discharges rather than by a direct effect of these agents. Furthermore, when the injections were repeated more often than once a week, a depression of the partially isolated regions of the nervous system developed first, which resulted in a reversal of the characteristic pattern of the convulsion. The investigations of Teasdall (1950), Teasdall and Stavraky (1950) and Seguin, Fretz and Stavraky (1957) referred to previously, showed that corpus-callotomized cats and rats were particularly prone to develop such a depression. Thus it was established that partial isolation renders neurones more excitable, but at the same time also more unstable in their responses, and predisposes them to periods of prolonged depression if they are exhausted by repeated activation. On the other hand, spacing the convulsions at appropriate intervals (once every seven to ten days) enhances the convulsability. This effect could be produced more easily in the partially denervated neurones than in the rest of the central nervous system. The restoration of extinct conditioned reflexes by means of induced convulsions observed by means of a delicate technique in intact animals (Kessler and Gellhorn, 1943; Gellhorn and Minatoya, 1943) may be regarded as another example of this trend.

Thus it appears that the processes which go on in neurones subjected to repeated discharges of a convulsive type are similar in the intact nervous system and in partially denervated regions. In the latter, however, the changes, whether of excitation or inhibition, are brought about more quickly and are more profound. It is suggested that the beneficial effects of convulsive therapy depend on the fact that in disease there exists a population of supersensitive neurones the excitability of which will be altered by repeated convulsions more quickly and to a greater extent than that of the intact nerve cells, and that their activity will be thus brought into line with that of the rest of the nervous system. Whatever the intimate nature of the pathological process may be, the knowledge of the behaviour of supersensitive nerve cells when they are exposed to repeated, excessive stimulation may lead to a more physiological approach to convulsive therapy.

CHAPTER III

Deafferentation and Supersensitivity

1. Effects of Stimulating and Depressing Chemical Agents on Deafferented Spinal Neurones

When the results of section of descending pathways of the central nervous system are considered, one inevitably recalls that, according to the widely known concept of Hughlings Jackson (1884), a destructive lesion which affects the higher levels of the central nervous system causes, besides a loss of control of the higher over the lower levels, a "release" or "disinhibition" of the lower levels manifesting itself in an increased activity of these levels. Head (1921) expanded Jackson's concepts of release of function in the nervous system to include the sensory sphere. He suggested that the exaggerated responses to sensory impressions which occur in the thalamic syndrome are also manifestations of a release from cerebral cortical control, and contrasted this with the loss of sensation which results from injury to the sensory paths in the cord. Furthermore, Magoun and co-workers (Magoun, 1944, 1950, 1953 a, b; Magoun and Rhines, 1947; Lindsley, 1952) drew attention to the facilitatory and inhibitory effects exerted by the bulbar reticular formation on spinal motor neurones and to the interdependence of this mechanism and the cerebral cortical projections. As long as one deals with cerebral ablations or with lesions of the descending pathways in the spinal cord, it is possible to account for the exaggeration of reflex adaptations and even possibly for the increased effectiveness of chemical stimulating agents on partially isolated neurones (described in the preceding chapters) by assuming that their overaction is due to the phenomenon of release and to an augmented influx from unopposed facilitatory mechanisms. However, the delay in the appearance and the progressive nature of the exaggerated sensitivity in the denervated regions contrasted with the immediate changes occurring after a destructive lesion in the central nervous system seemed to indicate that a gradual change in some basic property of the denervated cells was taking place along with the immediate effect caused by the interruption of anatomical connections. In order to ascertain whether an increase in sensitivity of the spinal neurones takes place following their partial isolation it seemed important to find out whether the principle of sensitization by denervation holds good only in regard to the influence of higher over lower levels in the central nervous system or whether it can also occur within a reflex arc.

Several experimental approaches to this problem will be considered. In one series of experiments stimulating and depressing chemical agents were tested on spinal neurones isolated by means of section of the posterior nerve roots. This study was carried out by Drake and Stavraky (1948 b) with a technique closely resembling that described in chapter II. In cats the left hind limb was deafferented by aseptic, intradural section of L2 to S2 dorsal roots, and after various periods of time ranging from several hours to three months, the brain and medulla were pithed under ether anaesthesia and the contractions of both quadriceps muscles were recorded. Care was taken to maintain the temperature, ventilation and other conditions at an optimal level and to allow for complete recovery from the effects of the initial anaesthesia before beginning the experiment; this was judged to be adequate when a spread of the knee jerk from the control to the deafferented side was observed. Acetylcholine, strychnine, pentylenetetrazol, camphor and adrenaline were injected intra-aortally to test the sensitivity of the motor neurones on the two sides of the spinal cord. Short periods of asphyxia were also employed. As an example of a depressing drug sodium pentobarbital was chosen.

As shown in Fig. 17, shortly after deafferentation appropriate quantities of the stimulating agents caused smaller contractions of the quadriceps muscle on the side on which the dorsal roots were severed than those on the intact side. This was true of all the drugs tested. These effects had been described previously in the case of strychnine and had led to the belief that strychnine acted on the afferent part of the reflex arc (Baglioni, 1900; Moore and Oertel, 1899; Pike, Coombs and Hastings, 1919; Verworn, 1900). However, this view was not substantiated by Dusser de Barenne (1910 a, b, c) in his work on the dermatomes. Later he and his co-workers (Dusser de Barenne and McCulloch, 1939; Dusser de Barenne, Garol and McCulloch, 1942; Ward and Kennard, 1942) brought forth a new concept according to which strychnine acted on the motor or intercalated neurones as a cholinergic stimulant, possibly as an anti-esterase substance. On the other hand, Sherrington (1905 b, 1906) held that it acted by converting inhibition within the central nervous system into excitation or that it depressed

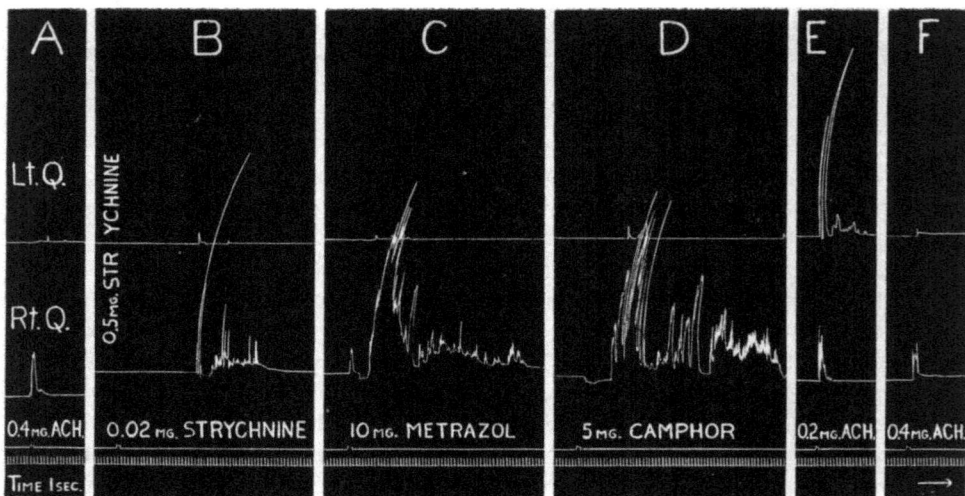

Fig. 17. Suppression of irritability of spinal neurones 6 hours after section of left L 2 to S 2 dorsal nerve roots. Spinal cat: isotonic myograms of left and right quadriceps (Lt.Q. and Rt.Q.); all drugs in this and next two figures injected intra-aortally. Note reversal of effect of acetylcholine in tracing *E* as compared with *A*, following injection of 0.5 mgm. strychnine, and return of initial response one hour later (*F*). (Drake and Stavraky, 1948)

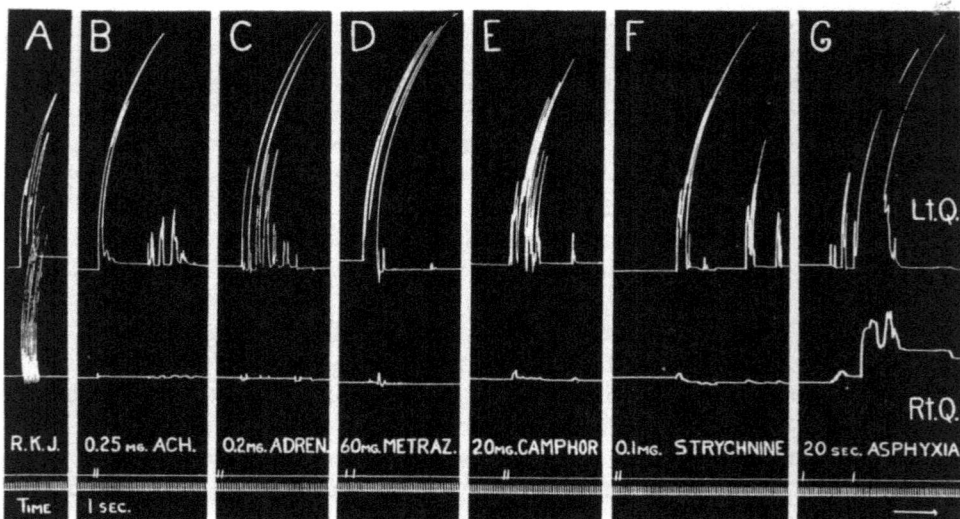

Fig. 18. Sensitization of spinal neurones eight days after section of L 2 to S 2 dorsal nerve roots in a similar experiment to one shown in Fig. 17. Note in *A* tapping of the right patellar ligament causes a bilateral response, the contractions of the quadriceps being more prolonged on the deafferented side. (Drake and Stavraky, 1948)

the inhibitory processes. This latter view is upheld by Eccles and his co-workers (Bradley and Schlapp, 1950; Eccles, 1953, pp. 169, 170; Bradley and Eccles, 1953; Bradley, Easton and Eccles, 1953) in relation to direct inhibition tested in monosynaptic reflexes.

When the time interval between deafferentation and the actual experiment was extended, the relative sensitivity of the two sides of the spinal cord to chemical stimuli gradually reversed itself: more pronounced responses being elicited from the deafferented side (Fig. 18). This effect could be demonstrated under special experimental conditions as early as six to eighteen hours following deafferentation but became prominent several weeks after section of the dorsal roots.

The quantities of the injected drugs had to be kept small in order not to cause a prolonged maximal response on both sides. When the response was maximal, the greater sensitivity of the deafferented side was still evident because the contractions started first on that side. However, when excessive quantities of the stimulating agents produced a prolonged series of maximal contractions, the deafferented side was seen to fatigue first and thus an apparent reversal of the effect could be brought about.

The deafferented spinal neurones were found to be more sensitive not only to stimulating agents and possibly to fatigue and other adverse conditions, but also to the action of sodium pentobarbital. Fig. 19 shows that an injection of sodium pentobarbital (0.58 mgm./kgm.) left the intact neurones practically unaffected but greatly depressed the deafferented neurones and thus reversed the effects of acetylcholine. This action of sodium pentobarbital was of relatively short duration, lasting from several minutes to half an hour. Successive injections of strychnine seemed to accelerate restoration of the sensitivity of the deafferented neurones when they were depressed by barbiturates, this being similar to the effect of strychnine on the action of acetylcholine in the early stages after deafferentation (Fig. 17 E).

In order to equalize as much as possible the experimental conditions on both sides of the spinal cord and to exclude any reflex influences coming from the intact side, the control side of some animals was deafferented at the time of the experiment. This second deafferentation depressed the irritability of the control side and thus further increased the difference between the responses of the two quadriceps. In other experiments greater contractions of the quadriceps on the deafferented side were obtained after clamping the abdominal aorta at the bifurcation to exclude any direct effect of the chemical agents on the muscles. Section of the femoral nerves completely abolished the effects of all the drugs

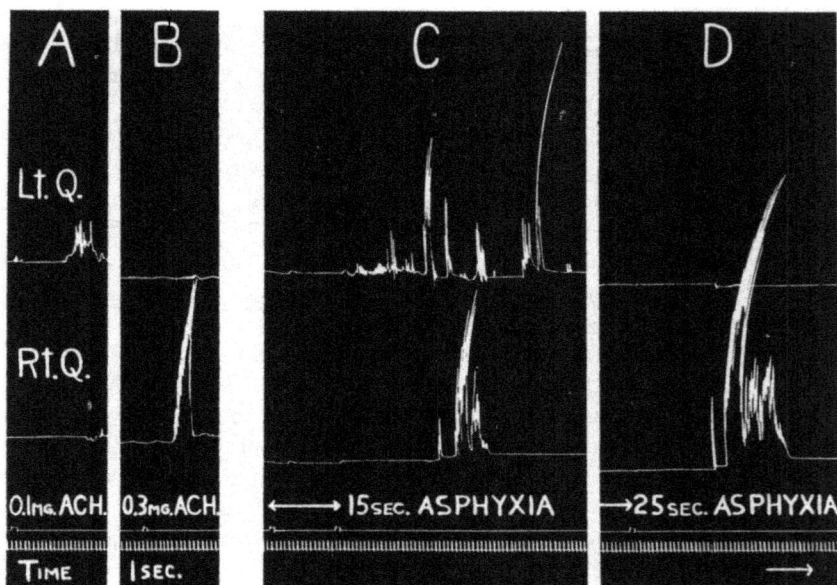

FIG. 19. Selective depression of the deafferented spinal neurones by 0.58 mgm./kgm. of pentobarbital in a spinal cat (left L 2 to S 2 dorsal roots severed sixty-four days before experiment). Tracings *A* and *C* show an increased sensitivity of the deafferented side to acetylcholine and asphyxia (ten seconds of asphyxia not shown in the figure gave a selective response on the deafferented side only). After the injection of pentobarbital, in tracings *B* and *D*, a larger quantity of acetylcholine and a longer period of asphyxia stimulated the spinal neurones on the control side in the absence of a response from the deafferented neurones. (Drake and Stavraky, 1948)

except acetylcholine: the latter was seen to stimulate the quadriceps directly in the highest range of doses used.

In another set of experiments aseptic deafferentation of one fore limb was carried out in cats by intradural section of C3 to T2 dorsal roots. Ranson (1928) and Ranson and Hinsey (1929) noted in the deafferented hind limb a tendency to hyperextension. The fore limb, on the other hand, though extended at the elbow, showed a tendency to flexion at the wrist, the animal stepping often on the back of the paw and developing ulcers in this region. Initial hypotonia, ataxic gait and loss of the sense of position were present in both the hind and fore limbs after deafferentation.

When acetylcholine, camphor or pentylenetetrazol was injected into these animals at seven- to ten-day intervals, an asymmetrical response developed and persisted throughout the period of observation, which

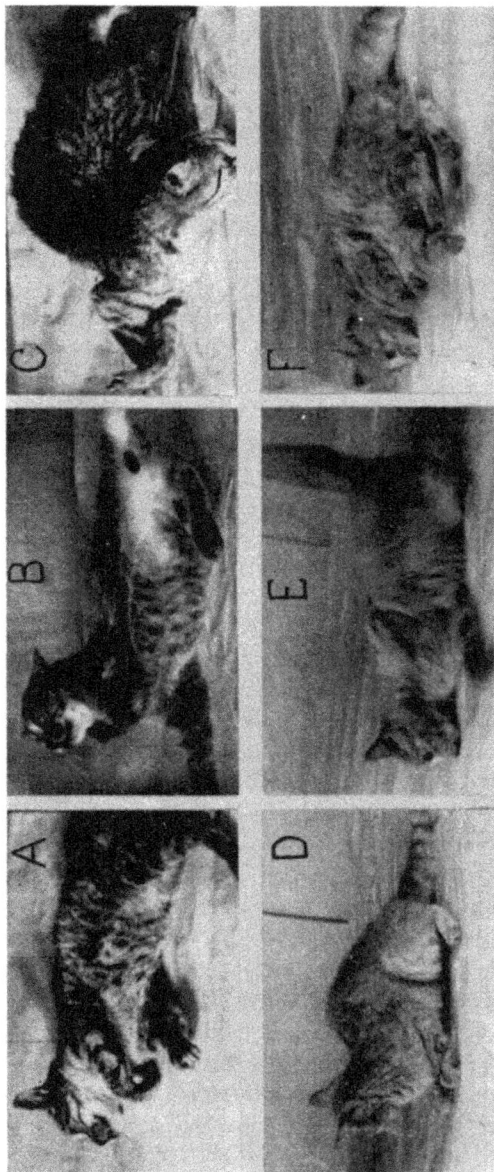

FIG. 20. Moving picture recordings of convulsions induced by acetylcholine and camphor in two cats with deafferented forelimbs; left C 3 to T 2 dorsal nerve roots severed six and four months before experiment respectively.

A, B, and C show the effect of intravenous injection of 0.2 mgm. of acetylcholine per kgm. of body weight. A: twenty to thirty seconds after injection. Initial tonic phase of the convulsion: deafferented forelimb is flexed, all other extremities extended. B: seventy-five seconds after injection. Generalized convulsion is over: animal attempts to get up with left front paw still rigidly flexed. C: ninety seconds after injection. Animal actually gets up, but is unable to maintain its balance on three legs owing to persisting contraction of the muscles of the left fore limb and falls over to that side.

D, E, and F show effect of intravenous injection of 4.8 mgm. of camphor monobromide per kgm. body weight. D: posture of cat before and immediately following the injection. E: fifteen to twenty seconds after the injection. Sharp generalized convulsion with left fore limb in flexion. F: sixty to ninety seconds after the injection. Generalized convulsion subsides but flexion of deafferented left fore limb persists accompanied by some crossed extension of the intact right fore limb. (Drake and Stavraky, 1948)

lasted up to six months. Acetylcholine produced characteristic tonic convulsions which often terminated in a clonic phase. At the onset of the convulsion the deafferented limb usually stiffened first and made a wide sweep, the animal then falling on one side with the intact limbs rigidly extended, the deafferented one becoming flexed and drawn up in front of the neck under the lower jaw (Fig. 20 A, B). Occasionally, if the paw got past the lower jaw, it was flexed over the head so that the protruding claws were pressed into the skin behind the ear, or the paw made a wide sweep, resembling a grasping movement, which was terminated by the claws being sunk into the table in front of the animal. Regardless of the final position of the limb, it stayed tense for fifteen to thirty seconds after the generalized tonic convulsion was over, the animal at times making an attempt to get up with the deafferented paw rigidly flexed (Fig. 20 B, C). This often resulted in falling to the side of the deafferented extremity. When the paw finally relaxed, it did so rapidly and completely without the period of hypertonicity which was characteristic of the limbs of cats with cerebral ablations (Stavraky, 1943; Drake, Seguin and Stavraky, 1956).

Pentylenetetrazol and camphor monobromate also caused asymmetrical generalized convulsions. At the height of the tonic phase both fore limbs were usually flexed and drawn up to, or over, the head, which became flexed on the neck. The deafferented fore limb underwent flexion first and was withdrawn farther over the head, the contraction in flexion persisting longer in it than in the intact limb. Injections of the same drugs into cats with deafferented hind limbs produced less asymmetry of movement during and following a convulsion, but the contractions, again, first began in the deafferented extremities, often outlasted those in other limbs, and seemed to be more powerful.

Thus the most characteristic feature of the convulsion was the exaggerated flexion of the deafferented extremities. As described by Sherrington (1910) and Miller (1931) strong stimulation of afferent nerves in decerebrate cats causes flexion of the extremities on the same side of the body, and extension contralaterally. It seems that in the experiments of Drake and Stavraky (1948 b) a sensitization of the neurones responsible for the flexion caused these deafferented neurones to respond to chemical stimuli in the same manner as they had to afferent impulses. This evoked something like an "after-image" of the normal response to a nociceptive stimulus.

In a study of the role which the cells of the dorsal root ganglion play in the sensitivity of the deafferented spinal neurones, Luco and Eyzaguirre (1952) confirmed the basic findings of Drake and Stavraky

(1948 b). At different intervals after section of the dorsal nerve roots anywhere between L7 and S4, the cats were curarized and made spinal. Electrical recordings of the activity of the "anterior horn cells" made from the central ends of the cut anterior nerve roots showed that section of a dorsal nerve root six to seventy-five days before the experiment intensified the discharges of the corresponding motor neurones when the latter were stimulated by intravenous injections of strychnine or pentylenetetrazol, or by asphyxia. This was true when the nerves were severed either centrally or peripherally to the dorsal root ganglion, but the effect of the proximal section was much more pronounced.

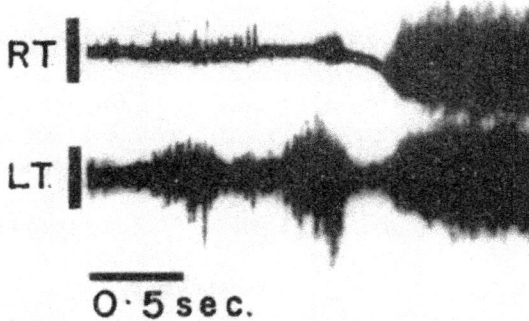

O·5 sec.

FIG. 21. Influence of the dorsal root ganglion on sensitivity of spinal neurones. Oscilloscopic tracings of the response to 0.15 mgm. strychnine (I.V.) recorded from corresponding ventral nerve roots thirty-two days after bilateral section of S 1 dorsal roots. Right (Rt.) dorsal root was severed peripherally to the ganglion: left (Lt.) dorsal root severed between the ganglion and the spinal cord (Calibration: 50 μV.). Note greater initial response to strychnine, which was injected shortly before the beginning of the record, on the left side. Similar results were obtained with asphyxia and pentylenetetrazol. (Luco and Eyzaguirre, 1952)

Particularly interesting were the experiments in which the dorsal roots were severed on two sides simultaneously. As shown in Fig. 21 strychnine caused a more prompt and greater response on the side on which the ganglion had been isolated from the spinal cord. This effect of strychnine took place in spite of the fact that the interruption of afferent impulses to the spinal neurones at the S1 level on both sides was completely symmetrical in time and distribution. Furthermore, curarization of the animal precluded contraction of the muscles after the injection of strychnine, thus excluding any proprioceptive impulses from influencing the motor neurones via neighbouring segments.

Luco and Eyzaguirre (1952) found that the augmented excitability required several days to develop, but they failed to see any initial depression after section of the sensory nerve root. However, as only one dorsal root on each side had been severed, this result is not unexpected and is in keeping with the observations of Mott and Sherrington (1894) who found an absence of any change in motor activity in monkeys after similar procedures, whereas deafferentation of a complete limb resulted in a gross impairment of movements. An overlap of the sensory termina-tions in the spinal cord was assumed to have been the responsible factor for this difference and probably could account also for the lack of initial depression in the experiments of Luco and Eyzaguirre (1952).

Thus the experiments of Drake and Stavraky (1948 b) and of Luco and Eyzaguirre (1952) show that in completely equal reflexological conditions the two sides of the spinal cord may differ from each other in their responses to chemical stimulating or depressing agents, the effect depending on the duration and nature of the deafferentation on each side. In the instance of bilateral deafferentation, the side on which the dorsal roots were severed shortly before the experiment was generally less sensitive than the one on which the afferent roots were severed some time previously. On the other hand, a section of the nerve root centrally to the ganglion resulted in a greater excitability of the spinal neurones than did a section below the ganglion. This latter finding is in agreement with Cannon's (1939) postulate that the supersensitivity which develops in penultimately isolated structures is less pronounced than the one seen after direct denervation. The contention that chronically deafferented neurones are supersensitive instead of being less depressed than the ones on the side of the acute radicotomy is illustrated by the fact that chronically deafferented neurones are more sensitive to chemical agents than are the neurones on the control side in which the dorsal nerve roots were left intact. The possibility that under these experimental conditions the effect is due to the autogenetic inhibitory influence exerted by the muscle and tendon proprioceptors, which limit the contractions of the muscle on the control side, while the one on the deafferented side is released from this influence, is untenable in view of the following con-siderations. (1) In experiments with electrical recordings, the animals were curarized and the nerve roots severed bilaterally, thus the effect of the chemical stimulating agents was determined in the absence of any muscular contractions or connections. (2) The greater effectiveness of depressing agents such as pentobarbital on the side on which the dorsal nerve roots were severed cannot be attributed to any peripheral effect.

2. RESPONSES OF DEAFFERENTED SPINAL NEURONES TO IMPULSES REACHING THEM BY WAY OF THE CORTICO-SPINAL TRACTS

If section of the posterior nerve roots leads to sensitization of the corresponding spinal centres to chemical agents, what would be the response of the deafferented spinal neurones to impulses which reach them by way of the cortico-spinal tracts? A study of this was undertaken by Teasdall and Stavraky (1953) in cats after deafferentation of one hind limb. At various times after the operation the animals were decerebrated by a transcollicular section, suspended by means of a sling, and the movements of the hind limbs were recorded. In these experiments the head was maintained in a fixed position with the labial cleft inclined at forty-five degrees below the horizontal plane, and the basis pedunculi was stimulated alternately on each side, the intact limb serving as a control for the deafferented one. The position of the electrodes was marked by means of bristles, and at the end of the experiment the brain and spinal cord were removed, fixed, and the exact point of stimulation as well as the completeness of the deafferentation were determined histologically.

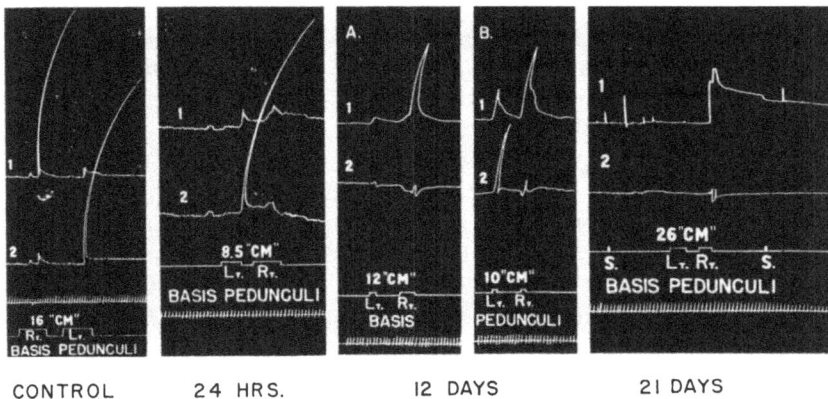

CONTROL   24 HRS.   12 DAYS   21 DAYS

FIG. 22. Flexor responses evoked by electric stimulation of basis pedunculi in decerebrate cat after section of L 3 to S 3 left dorsal nerve roots. (Tracings 1 and 2 are left and right hind limb respectively; time 0.5 second; coil distance in centimetres.) Note equal responses in a control experiment; depression of the response on the deafferented side twenty-four hours after section, and a progressively greater sensitivity of the deafferented side twelve and twenty-one days after section which is expressed in a reduction of the threshold stimulus, a prolongation of the after-effect, spread to the opposite side and, in the last recording, a selective sensitivity to thermal stimulation of the cut surface of the brain stem induced by irrigation with warm saline (S). (Teasdall and Stavraky, 1953)

Immediately after unilateral section of the dorsal nerve roots no response to stimulation was observed in either the acutely deafferented or the intact limb even when excessively strong currents were employed. It should be noted, however, that with the application of single square-wave stimuli to the pyramid Bülbring *et al.* (1948) have obtained movements in acutely deafferented hind limbs of decerebrate cats. The depression seen following deafferentation is, therefore, not absolute and may be overcome by the use of appropriate stimuli.

Twenty-four hours after the operation a flexor response was consistently observed in the intact hind limb, while with identical stimuli there was only a minimal response on the side of the deafferentation (Fig. 22). A stronger current, however, would produce a flexor response in the deafferented hind limb.

Two to four days after section flexor responses were elicited in both the intact and the deafferented hind limbs, the amplitude of the contractions being equal on both sides. Strengths of current required to produce the responses were also identical on the two sides, but were somewhat larger than in normal decerebrate preparations.

Five to twenty days after deafferentation a current too weak to produce a response in the intact limb regularly produced flexion of the deafferented limb. Higher voltages elicited contractions of greater amplitude on the deafferentated side. Bilateral responses were also observed following application of relatively weak currents to either basis pedunculi, the ipsilateral flexion being more marked on the side of the deafferentation. Even greater differences were observed between the responses of the two limbs twenty-one to forty-seven days after deafferentation. In addition to the lower threshold and greater amplitude of flexion, the responses of the deafferented limb were often maintained for five to six seconds following cessation of the stimulus. This prolonged after-discharge was characteristic of the deafferented neurones and was not observed on the intact side. At this stage the deafferented neurones were more sensitive, not only to electrical stimulation of the corresponding basis pedunculi, but also to thermal stimulation of this structure: following the application of warm saline to the cut surface of the brain, a flexor response was often noted in the deafferented limb but not on the intact side. Furthermore, though with the technique employed it was impossible to determine accurately the time interval between the application of the stimulus and the onset of the contraction, it was felt that the latent period was considerably shorter for the deafferented than for the intact limb.

Under certain conditions, such as fatigue following repeated stimulation, the spinal centres on the deafferented side became depressed. A similar depression occurred during a lowering of the body temperature or with general deterioration of the preparation.

Many previous investigators (Thiele, 1905; Graham Brown, 1913, 1915; Weed, 1914; Hinsey, Ranson and McNattin, 1930; Hinsey, Ranson and Dixon, 1930; and others) stimulated the cut surface of the mesencephalon with a technique similar to that employed by Teasdall and Stavraky (1953). In general, electric stimulation of the cortico-spinal tract in the region of the basis pedunculi produced a rapid flexion of the contralateral fore and hind limbs which was followed by a prompt relaxation on cessation of the stimulus. On the other hand, stimulation of the mesencephalic tegmentum resulted in slow movements of all four extremities and in a curvature of the spine, the concavity of which was directed to the side of the stimulation. These movements persisted for several seconds after the cessation of the stimulus and could be distinguished from the effects of stimulation of the cortico-spinal tract.

In accordance with these findings, the tendency towards a bilateral response and the prolonged after-discharge of the deafferented spinal neurones found by Teasdall and Stavraky (1953) may be attributed to the stimulation of the mesencephalic tegmentum. Although this possibility cannot be ruled out altogether, it should be emphasized that the electrodes were placed directly on the basis pedunculi, this being checked by subsequent histological examination. Consequently, the stimulation of structures in the tegmental area could be due only to spread of current from the region of the basis pedunculi and in control experiments it was found that considerably stronger currents were required to produce a generalized response of this type. Also the absence of the curvature of the body and of fore limb responses and the appearance only in the deafferented hind limb of the prolonged after-contraction make the possibility of a spread of current to the tegmentum most unlikely. Thus the prolonged after-discharge may be attributed to the supersensitivity of the deafferented spinal neurones to impulses reaching them via the cortico-spinal tracts.

If the basis pedunculi was stimulated electrically or thermally during the phase of the after-discharge (Fig. 22), the deafferented extremity rapidly assumed its resting attitude of extension. This inhibition of the flexor contractions occurred only when the deafferented spinal neurones were already in activity at the time of stimulation of the cortico-spinal tract. An inhibition which results from an apparent overexcitation will be the topic of chapter IV, but it should be emphasized at this time that inhibition was consistently observed in widely diverse experimental situations and became almost an expected part of the maximal state of sensitivity.

The appearance of these exaggerated and modified responses in the deafferented extremity to electrical stimulation of the cerebral peduncle coincided with the post-operative period in which spontaneous movements became apparent in the deafferented limb. At this time the cat was capable of initiating movements of progression in the deafferented limb which were synchronous with movements of the other extremities. However, during locomotion the deafferented limb was flexed to a greater degree and then extended forcefully, being poorly placed in contact with the ground. This gross inco-ordination resembled clinical ataxia in many ways.

The effect of stimulation of the motor cortex on deafferented spinal neurones was studied by Sherrington (1893) in a monkey anaesthetized with alcohol-ether-chloroform; section of L5 to S2 dorsal nerve roots did not diminish the effectiveness of cortical stimulations. In fact, the threshold of the corresponding motor cortex to electrical stimulation was lowered by the deafferentation. Teasdall (1950) confirmed Sherrington's observation on four cats following section of L3 to S3 dorsal roots. Sherrington exposed only one motor cortex and studied the response of the corresponding deafferented hind limb to its stimulation. Under these conditions he attributed the result to the diminished tone of the deafferented limb. In Teasdall's (1950) experiments both motor cortices were exposed and stimulated alternately and the threshold was found to be lowered on both sides by a one-sided deafferentation. No simple explanation can be offered for the effect of acute deafferentation on the responses of the opposite limb. Some augmentory effects of unilateral section of the dorsal roots on the opposite side of the spinal cord were seen also by Terzian and Terzuolo (1954). On the other hand, the experiments of Smith, Mettler and Culler (1940) show clearly the unilateral component of deafferentation. A rhythmic phasic response was induced by these investigators in the foreleg of cats under ether anaesthesia by appropriate continuous electric stimulation of the motor cortex. After section of C5-T1 dorsal roots, the cortical stimulation caused full tonic flexion which was followed by a relaxation in spite of a continued stimulation; stimulation of the opposite motor cortex yielded the usual phasic response. Smith, Mettler and Culler (1940) concluded that the phasic response was dependent, in part, upon the inhibitory proprioceptive impulses set up in the contracting muscles, and a further analysis revealed that the cerebellum also contributed to this response. Another observation may be relevant: Asratian (1953, p. 355), studying the recovery of function in chronic dogs after section of the dorsal columns of the spinal cord, noted a temporary depression of motor activity below the level of the section. At certain times after the operation, when the depressed state of the segmental reflex was still present, voluntary movements and movements brought about by application of sensory stimuli to the head of the animal were found to be brisk and powerful and were not as rapidly fatigued as the segmental reflex adaptations.

3. REFLEX ADAPTATIONS AND TONE IN DEAFFERENTED SPINAL NEURONES

From the preceding observations it may be concluded that interruption of afferent connections leads to an alteration of the responses of spinal neurones' to chemical stimulation and to nerve impulses reaching them by way of the cortico-spinal tracts. These changes are of two types, one immediate and the other delayed. The latter develops gradually under most experimental conditions and seems to depend upon an alteration of the state of the deafferented nerve cells rather than upon a sudden interruption of specific anatomical connections. If this be the case, then reflexes which are not abolished by the interruption of the dorsal nerve roots should be affected in a similar manner.

The permanent abolition of the intrinsic reflexes in a deafferented limb excludes the possibility of studying the responses induced by stretch of the muscles or of the flexor withdrawal brought about by ipsilateral nociceptive stimulation. Sherrington (1893) made some interesting observations while mapping out the "remaining sensibility" after section of one root above and one below the root to be studied. In such animals the area of remaining cutaneous sensation was observed to increase progressively in size after the operation. According to Sherrington (1893), the initial restriction of the area of reflex response was due to "local spinal depression" as judged by the long latency and enfeebled responses to cutaneous stimulation. This condition was gradually supplanted by a "local spinal excitation" and extension of the field of cutaneous sensibility. The initial depression could be shortened by repeated application of the stimulus even when the latter was insufficient to evoke obvious responses. The extrinsic reflexes, such as the crossed extensor reflex and the Philippson's Reflex (1905) at the segmental level, the scratch reflex in the intersegmental field, and the more complex adaptations, such as the linear acceleratory and tonic labyrinthine and neck reflexes, were also extensively studied. The part played by many of these reflexes in muscular tone also received much attention. Consequently, the study of the effects of deafferentation on individual reflexes and on tone in general may be considered simultaneously.

Brondgeest (1860) and Mommsen (1885) noted that tone in the limbs of frogs was diminished by dorsal root section. Subsequently, the difference between the responses of the normal and deafferented quadriceps muscle in the decerebrate cat was described by Sherrington (1898 a, b, 1909). The intact muscle always developed decerebrate rigidity, showed an exaggerated tendon jerk, and exhibited prominent lengthening and shortening reactions. The acutely deafferented quadriceps revealed none of these phenomena which Sherrington believed to be characteristic of the hypertonic state. Accordingly, the proprioceptive reflex theory of muscle tone was based on the fact that tone in decerebrate animals was primarily a stretch reflex dependent upon the afferent nerve supply of the extensor muscles (Liddell and Sherrington, 1924, 1925; Fulton and Liddell, 1925; Fulton, 1926).

Although decerebrate rigidity was abolished by dorsal root section, the deafferented quadriceps was found by Sherrington (1909) to participate in other postural adaptations such as the crossed extensor reflex. This reflex as studied in the vastocrureus muscle was found to be actually exaggerated by deafferentation, a fact attributed by Sherrington (1909) to the resulting hypotonia. On the other hand, from the observations of Fulton and Liddell (1925) on the crossed extensor reflex in intact and deafferented muscles and from the analysis of Fulton and Pi-Suñer (1928) of the lengthening reaction, a concept was developed which postulated that sensory nerve fibres originating in muscle spindles make automatic adjustments in the spinal cord which lead to greater smoothness and precision of contraction and prevent abrupt changes of tension which might injure the muscle.

Much of the recent electrophysiological analysis of the action of muscle and tendon proprioceptors bears out, in a general way, the importance of this mechanism in the regulation of motor activity. The studies of Adrian (1928, 1931, 1932), Adrian and Bronk (1929), Adrian and Umrath (1929), Adrian and Zotterman (1926 a, b), Matthews (1931 a, b, 1933), Lloyd (1943 a, b, c, 1944, 1946 a, b), Laporte and Lloyd (1952), Leksell (1945), Hunt (1951, 1952 a), Hunt and Kuffler (1951 a, b), Kuffler and Hunt (1952), Kuffler, Hunt and Quilliam (1951), Granit (1950, 1955), Granit and Kaada (1952), McCouch *et al.* (1950, 1951) and many others summarized by Bremer (1953), Kuffler (1953), Kaada (1953), Magoun (1953 b) and Moruzzi (1953) and fully reviewed by Granit (1955, chapters VI, VII, pp. 191–276) indicate the great complexity of this peripheral sensory contribution to reflex control of muscular movement which, as stated by Kuffler (1953), "consists in a continuous dual excitatory-inhibitory action, changing with stretch and movement." However, from the studies of tonic adaptations, it became apparent that a number of non-myotatic reflexes participate in the co-ordination of muscular activity. Particularly prominent among them were labyrinthine and neck reflex influences. This had led Sherrington (1915) and Magnus (1924, 1926) to the conclusion that, while tone of a given muscle depended essentially upon the integrity of the dorsal nerve roots, it was not wholly independent of afferent impulses coming from other levels. Impressed by the fact that section of the VIIIth cranial nerves did not abolish decerebrate rigidity while dorsal radicotomy did, Sherrington (1915) at first attributed only a secondary role to non-segmental mechanisms in the maintenance of tone. However, when Foerster (1918) attempted to relieve spasticity in human subjects by surgical section of the dorsal roots, the operation was not a success (Steinke, 1918). On the other hand, Magnus (1926) emphasized the importance of these mechanisms. In collaboration with Liljestrand (1919), he observed that section of the spinal cord in the region of the 12th thoracic segment restored the decerebrate rigidity in the fore limbs made flaccid by deafferentation (bilateral section of C6–T2 dorsal roots). These experiments laid the groundwork for the subsequent analysis of non-segmental factors which influence muscular tone. Furthermore, Liljestrand and Magnus (1919) found that during the first few days after section of the dorsal nerve roots of one brachial plexus, the triceps muscle was flaccid but retained its capacity for active movement. After about one week the muscle again acquired its tone; this gradually increased with time and, if the animal was decerebrated at this stage, rigidity developed in the deafferented limb. This was confirmed by Pollock and Davis (1923, 1924, 1926, 1927 a, b, 1929, 1930 a, b, 1931) who noted that the rigidity in the deafferented foreleg was largely dependent on labyrinthine and neck reflexes. Particularly striking results were obtained by these latter investigators using the method of anaemic decerebration.

A subsequent analysis of the experiments of Pollock and Davis by Stella (1944 a, b, c, 1946) and the investigations of Cardin (1946 a, b, c) and Terzian and Terzuolo (1951) and Terzuolo and Terzian (1951, 1953) revealed that the anaemic method of decerebration removes the anterior lobe of the cerebellum, abolishing the cerebellar influences on the vestibular nuclei and thus inducing an intense decerebrate rigidity not only in a chronically but also in an acutely deafferented limb. As a result of this analysis, Stella (1944 a, b, c, 1946) described the different influences which participate in the maintenance of tone, distinguishing the myotatic, vestibular, cerebellar and spinal components. He omitted to take into consideration the influence of chronic deafferentation on the spinal neurones themselves apart from the effects of the exclusion of various anatomical regions of the central nervous system.

Yet the fact remained that while, on the whole, Coman (1928), Ranson (1928), Sprong (1929), Moldaver (1935) and many later investigators confirmed Brondgeest's (1860) and Sherrington's (1898 a, b, 1909) observations that a recently deafferented extremity was hypotonic, most of them found that "chronic" deafferentation had no such effect. A gradual return of tonus as postulated by Liljestrand and Magnus (1919) took place, and Sprong (1929) and Bremer (1932) concluded that, in these conditions, postural tonus was subject to sudden fluctuations and that a deafferented limb could be hypotonic or hypertonic depending on the conditions under which the observations were made.

In a more recent analysis Cardin (1951, 1952 a, b) and Terzian and Terzuolo (1954) showed that in chronic deafferentation there was present a "bulbar" tonic component not of a vestibular origin which was absent in acutely deafferented preparations. Thus Cardin (1951) found, in dogs subjected two months previously to post-brachial transection of the spinal cord and to section of the C1–D2 dorsal roots, that subsequent bilateral section of the VIIIth nerves did not prevent the appearance of decerebrate rigidity in the fore limbs after transcollicular section of the brain. In an analysis of this residual tone, Cardin (1952 a, b) found that two transections of the spinal cord at D12 and S3 levels, combined with bilateral section of L1 to S2 dorsal roots, resulted in an initial flaccidity of the hind limbs which slowly gave way to a "hypertonus." The latter reached its maximum in three or four weeks and was observed to last as long as ten months. From these experiments Cardin (1952 a, b) drew the conclusion that an intrinsic tone exists in the spinal centre. It should be noted, however, in regard to the latter conclusion, that Tower (1937), Tower, Bodian and Howe (1941) and Eccles (1941) in earlier experiments carried out along much the same lines as Cardin's analysis failed to observe any "autochthonous" activity in the isolated mammalian spinal cord. The discrepancy was not accounted for by Cardin, but support for the general idea that denervation of spinal motoneurones may result in their "spontaneous" activity was brought forth recently by Gelfan and Tarlov (1959). They showed that in dogs destruction of interneurones by asphyxiation results in rigidity, which develops regardless of a preceding section of the dorsal roots.

In a very exacting analysis of this condition, evidence was presented that lumbosacral motoneurones in such animals have lost the normal polysynaptic influence as a result of interneurone destruction. However, a considerably greater percentage of neurones in a "rigid motoneurone pool" responded to monosynaptic excitation than in a normal one and continuous motoneurone discharge persisted after deafferentation and decapitation of the animals. Small gamma motoneurones as well as large alpha ones were not silenced by these procedures. Gelfan and Tarlov concluded that denervation increases the excitability of moto-neurones to the point of discharging "spontaneously" and that the spontaneous discharges of spinal motoneurones were responsible for the enduring rigidity.

The importance of the time factor for the development of tone in deafferented extremities was fully recognized and subjected to an analysis by Bremer and his co-workers (Bremer, 1929, 1931, 1932; Moldaver, 1935, 1936; Bremer, Bonnet and Moldaver, 1942). Moldaver (1935, 1936) studied in a detailed manner, by means of isometric myographic and oscilloscopic recordings, the crossed extensor reflex in isolated muscles of decerebrate cats and of spinal toads and frogs. In the intact hind limb of the cat, following a short faradization of the contra-lateral sciatic nerve, this reflex was found to have the following charac-teristics: after a short latent period the contraction of the gastrocnemius-soleus-plantar muscle gradually reached a plateau which was maintained some ten seconds after the cessation of the stimulus and then returned slowly to its original level. The electrogram revealed an increased activity which corresponded to the development of the reflex contraction; the primary waves were synchronous with the electrical stimulus, with smaller secondary waves being interposed.

In an acutely deafferented extremity the crossed extensor reflex was considerably weakened. As seen in Fig. 23 B, one hour after section of the L5–S2 dorsal roots the latent period was lengthened, the ascent of the myogram was more gradual, and the contraction did not reach the height which it had attained before the section of the dorsal nerve roots. Furthermore, the relaxation of the muscle followed immediately the withdrawal of the stimulus. The electrogram paralleled the myogram, revealing a diminished recruitment of motor units which was expressed in a low amplitude of the potentials, the secondary waves being very markedly reduced. Moldaver (1935, 1936) noted that this effect was not due to a general depression of reflex activity because the corneal and pinnal reflexes remained unaffected by the experimental procedure, and the respiratory and cardiac rates were maintained within normal limits.

On the other hand, when the dorsal roots were sectioned three to

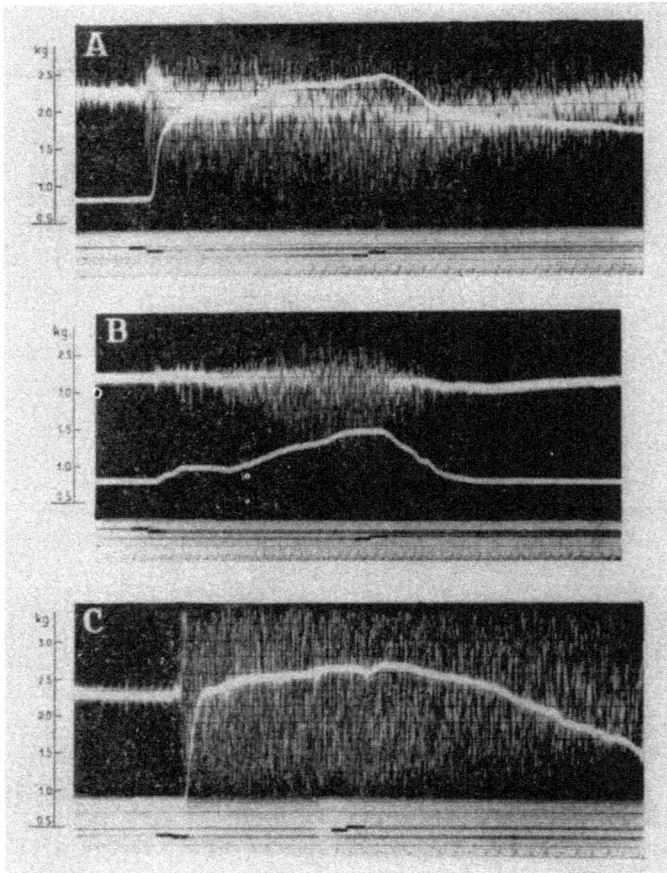

Fig. 23. Crossed extensor reflex in gastrocnemius-soleus-plantar muscle of decerebrate cat (oscillogram and isometric myogram; time 1/5 second). *A*: control. *B*: one hour after section of L 5 to S 2 dorsal nerve roots; shows depression. *C*: 8 days after deafferentation; shows augmentation of the response. (Moldaver, 1936)

twenty-one days previously, the crossed extensor reflex became markedly exaggerated (Fig. 23 C). The latent period was not noticeably modified, but the ascent of the myogram was brisk, resembling that of a nerve-muscle preparation. The plateau, thus readily attained, maintained itself for some time after cessation of stimulation, the muscle remaining in a contracted state up to five minutes. The electrogram was characterized by a large number of action potentials in which the rhythm was no longer discernible because the secondary waves were greatly augmented in amplitude and number.

Commenting on the exaggerated response seen in the chronically deafferented limb as opposed to the depressed state observed following acute severance of the dorsal roots, Moldaver (1936) concluded that it "appears to be more the expression of a profound modification of the excitability of the nervous centres following the interruption of the axis cylinders of the first afferent neurones" than of the effect of interruption of centripetal impulses coming from the muscle and tendon proprioceptors.

Bremer, Bonnet and Moldaver (1942) compared the after-discharge of the crossed extensor reflex in the deafferented hind limb of long duration to that observed in an intact extremity, and found it to be greatly increased in the former. They pointed out that the after-discharge of this reflex in the intact hind limb was of two types: true and false. The true after-discharge was a central phenomenon, while the false one resulted from proprioceptive nerve impulses originating in the muscle during the contraction. This latter reaction was abolished by dorsal root section. Bremer, Bonnet and Moldaver (1942) concluded that the "increased after-discharge of the crossed extensor reflex in the deafferented hind limb observed some days after the section of the dorsal nerve roots is a true exaggeration of the response and one of the manifestations of a neuronal hypersensitivity which follows deafferentation of nervous centres."

Furthermore, Bremer (1928) found that, when cats completely recovered from unilateral section of L4–S1 dorsal nerve roots, the deafferented hind limb exhibited a surprising hyperactivity compared to the opposite limb when vestibular reflexes were initiated by appropriate movement of the animal in space. While these reflexes were barely noticeable in normal animals and on the intact side of the operated cats, the slightest passive displacement was sufficient to set off the reaction in the deafferented leg. Thus, horizontal or vertical displacement of the animal provoked a brisk movement of the deafferented hind limb which was of great amplitude and resulted in active extension of the hip, knee and ankle when the animal was displaced from front to back, or downward. Displacement of the animal in the opposite direction resulted in active flexion of the same joints. Differential section of L4, 5, 6 dorsal roots resulted in a localized exaggeration of the reflexes in the muscles of the hip and thigh, while section of L7 and S1 dorsal roots had a similar effect in the muscles of the shank and foot. When procaine was injected in accordance with the technique of Liljestrand and Magnus (1919) into the quadriceps and gastrocnemius muscles in sufficient quantities to suppress the tone and tendon reflexes in these muscles, isometric myographic recordings showed no exaggeration of the crossed extensor reflex or of linear acceleratory reflexes. From this analysis the conclusion was drawn that the exaggeration of reflexes seen in the chronically deafferented limb resulted not from the elimination of the muscle and tendon receptors but was due to an augmented sensitivity of the spinal neurones following their partial isolation.

Terzian and Terzuolo (1951, 1954) and Terzuolo and Terzian (1951, 1953) reinvestigated the difference in tone and the responses to reflex stimulation in acutely and chronically deafferented limbs of decerebrate cats. They found that labyrinthine and neck reflexes as well as the facilitatory effects of electric stimulation of the cerebellum or of the brain-stem reticular formation were more prominent on the side of the chronic deafferentation. Thus, they confirmed and extended the observations of Bremer and his co-workers. Terzian and Terzuolo (1954) also found that chronic deafferentation of one side increased the excitability of the spinal motor neurones contralaterally but, of course, to a lesser degree. They discussed the results of their complex analysis in a most thought-provoking manner and concluded that the findings could be accounted for in the light of Cannon's (1939) Law of Denervation.

On the other hand, Eccles and McIntyre (1951, 1953) found that section of dorsal roots distal to the ganglion reduced the monosynaptic responses evoked, twenty-one to forty days later, by stimulation of the posterior root proximally to the ganglion, whereas polysynaptic discharges and those in the neighbouring segments were somewhat augmented. A brief burst along the afferent path restored the sensitivity of the monosynaptic channel, which remained enhanced for hours and influenced adjacent segments. The post-tetanic potentiation of the reflex evoked from disused roots ran an abnormally slow time-course, reaching a later summit and decaying much more slowly than on the control side. Strictly speaking, sectioning of dorsal roots distal to the ganglion did not denervate the spinal neurones; it only silenced them, and McIntyre (1953) felt that the depression of the monosynaptic responses "is at least in part related to the absence of the normal impulse barrage, and is not merely a sequel of the overall reaction of the primary afferent neurones to fibre amputation." However, even if it is granted that the excitability of the afferent neurones is not affected by the amputation of the distal fibres, the initial depression observed with a single-shock technique in the monosynaptic testing of the silenced neurones may be regarded as a special experimental situation and should not be compared too closely with results of chemical stimulation or with those obtained on repetitive stimulation. Also, the experiments of Eccles and McIntyre (1951, 1953) were carried out under pentobarbital anaesthesia. Drake and Stavraky (1948 b) and Seguin and Stavraky (1957) showed that barbiturates selectively depress the sensitivity of denervated spinal neurones; therefore, the true state of the basic excitability of the indirectly deafferented nerve cells is open to question in these very interesting experiments.

Post-tetanic potentiation fits in well with other observations of this nature such as the "stair-case" phenomenon seen in a stopped heart or the greater height of a tetanic contraction, as compared to the single twitch, in an isolated skeletal muscle. As McIntyre (1953) and Eccles (1953, pp. 203–27) point out, the demonstration of this phenomenon in the central nervous system may be important in the understanding of the process of learning. The slow time-course of the potentiation and its gradual decay are probably of great significance in the mechanism of supersensitivity of denervation and will be considered more fully later.

Thus it was gradually realized that crossed segmental and intersegmental spinal reflexes as well as labyrinthine and neck reflexes become exaggerated after chronic deafferentation of the appropriate neuronal pools. In the case of stato-tonic and stato-kinetic postural adjustments studied in survival experiments on non-decerebrated animals the emphasis was placed on the exaggeration of the linear acceleratory reflexes in deafferented extremities, while tonic labyrinthine and neck reflexes received relatively little attention. An investigation of these latter was undertaken by Teasdall (1950) and by Teasdall and Stavraky (1955) in cats, in which one fore limb was deafferented by means of intradural section of C3 to T3 dorsal roots. A similar operation involving L2 to S3 dorsal roots was done in the hind limb. At various periods of time following the operation the segmental, intersegmental, and suprasegmental reflexes were studied by direct visualization and by means of moving picture and other photographic recordings in both the deafferented and intact extremities. After an initial period of depression the crossed extensor reflex in the chronically deafferented limb was found to be greatly exaggerated. Intersegmental reflexes, particularly the scratch reflex, were also markedly increased in the deafferented limbs; so were the linear acceleratory reflexes. These findings were in agreement with the observations of Trendelenburg (1906), Bremer (1928), Ranson (1928), Sprong (1929), Pollock and Davis (1930 b), Moldaver (1935, 1936), Terzian and Terzuolo (1954) and others.

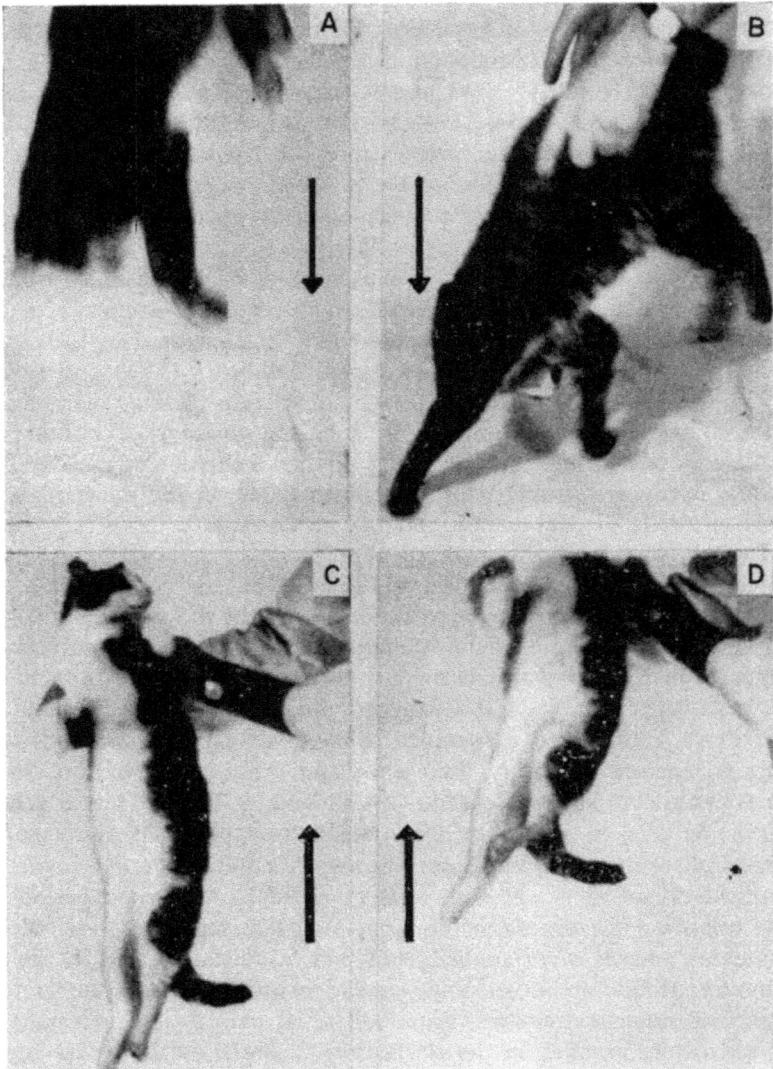

Fɪɢ. 24. Moving picture recordings of linear acceleratory reflexes in deafferented limbs of cats elicited in the vertical plane. *A* and *B*: cat held by pelvis and lowered to the ground (left C 3 to T 3 dorsal roots severed two months previously). Note semiflexed position of both fore limbs at beginning of descent (*A*), and the exaggerated extension of the deafferented limb when the animal reaches the ground (*B*). *C*. and *D*: cat held by upper part of the body and raised away from the ground (left L 2 to S 3 dorsal roots severed four months before the

With the re-establishment of these reflexes, tone reappeared in the deafferented extremity; furthermore, the alterations of tone in that limb seemed to be directly associated with these reflexes and therefore could be attributed largely to them. In the post-operative period the crossed extensor reflex and Philippson's reflex (1905) returned earlier in the deafferented hind limb than in the fore limb and they were more readily elicited in the former. Similarly, variations in muscle tone were noted sooner and were much greater in the hind limb than in the fore limb at corresponding times following deafferentation (one to seven days).

The deficiency of the crossed extensor reflex in the fore limb seemed to be compensated for by the overactivity of the linear acceleratory reflexes. They could be demonstrated in the fore limb within one week after section of the dorsal roots, a somewhat longer period of time after the operation (three to four weeks) being required before they were observed in the hind limb. These reflexes also influenced the tone in deafferented limb: for instance, if the cat was rapidly raised from the ground, extensor tone was decreased in the extremity, while the reverse occurred when the animal was lowered toward the ground (Fig. 24). These modifications of tone in the deafferented limb in response to linear acceleration were transient in nature and were evident only when there was an adequate stimulus. In this, they were similar to the tonic adjustments mediated by the crossed extensor reflex. Thus a considerable degree of muscle tone, transient in nature, was already evident in both the deafferented fore and hind limbs during the first month after section of the dorsal nerve roots.

The most striking feature of this study of reflex adjustments in the chronic experiment was, however, the change in the tonic labyrinthine and neck reflexes. Immediately after unilateral section of the dorsal nerve roots these reflexes were absent, not only in the deafferented limb, but also in the contralateral intact extremity. Within two weeks after section of the dorsal roots these reflex responses could be elicited in the intact limbs but not in the deafferented ones. One to two months after the operation the tonic neck and labyrinthine reflexes were once more present in the deafferented fore limb, while two or three months elapsed before the same reflexes could be elicited in the deafferented hind limb. When they became well established in the chronically deafferented limb, the tonic labyrinthine and neck reflexes showed a complete reversal from normal responses. Thus, if the animal was placed

---

test). Note almost symmetrical extension of both hind limbs at beginning of ascent (C) and the exaggerated flexion of the deafferented limb at height of ascent (D). (Teasdall and Stavraky, 1955)

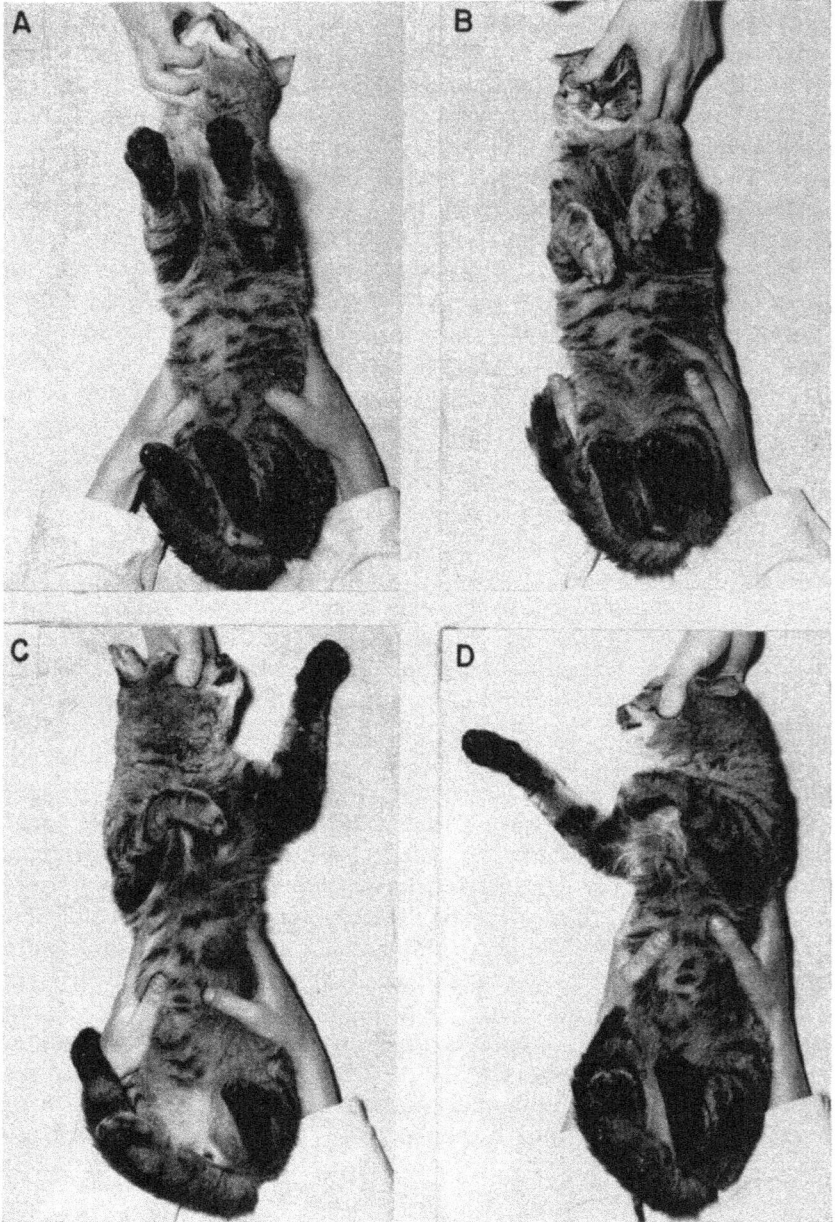

in the "maximal attitudinal extensor position" of Magnus (1924, 1926), that is, in the supine position with the labial cleft inclined at forty-five degrees above the horizontal plane, the intact fore limb was extended, while the deafferented one became semiflexed or completely flexed (Fig. 26 A). Ventroflexion of the head in this position resulted in an extension of the deafferented fore limb and flexion of the intact one (Fig. 26 B).

Similarly, when the animal was placed in the prone position with the labial cleft inclined at forty-five degrees below the horizontal plane— "minimal attitudinal position" of Magnus (1924, 1926)—the deafferented fore limb was rigidly extended, while the intact limb was flexed. Intermediate degrees of rotation in the vertical plane between the "minimal" and "maximal" attitudinal positions diminished the differences in posture between the deafferented and the intact fore limb; when the animal was midway between the two positions (that is, rotated through ninety degrees from either the "minimal" or "maximal" attitudinal position) the posture was identical in both fore limbs. Rotation of the head in the longitudinal axis also resulted in different responses of the intact and the deafferented extremities. Thus, when in the supine position the head was rotated to the left, this resulted in the usual flexion of the intact right or "vertex" limb and in a simultaneous flexion of the deafferented left or "jaw" limb, both extremities becoming relaxed (Fig. 26 C). On the other hand, rotation of the head to the right resulted in a simultaneous extension of both the intact right fore limb and of the left deafferented one (Fig. 26 D).

The reversal of the tonic labyrinthine and neck reflexes in the deafferented hind limb was best demonstrated by holding the animal by the scruff of the neck in the vertical plane. Ventroflexion of the head with the animal in this position resulted in the usual extension of the intact hind limb and flexion of the deafferented one, while dorsiflexion of the head resulted in flexion of the intact limb and extension of the deafferented limb (Fig. 27 C, D). The combination of bilateral labyrinthec-

---

FIG. 25. Tonic labyrinthine and neck reflexes in an intact, conscious cat in supine position. *A*: with labial cleft 45° above horizontal plane ("maximal extensor position" of Magnus). There is an increased extensor tone in both fore limbs. *B*: flexion of head causes reduction of tone in the forelegs. *C*: rotation of the head to the left causes extension of the left ("jaw") limb, and flexion of right ("skull") limb. *D*: rotation of the head to the right reverses the position of the limbs and the distribution of tone in them. Note absence of response in the hind limbs when the animal is in horizontal plane as compared to the vertical plane shown in Fig. 27. (Teasdall and Stavraky, 1955)

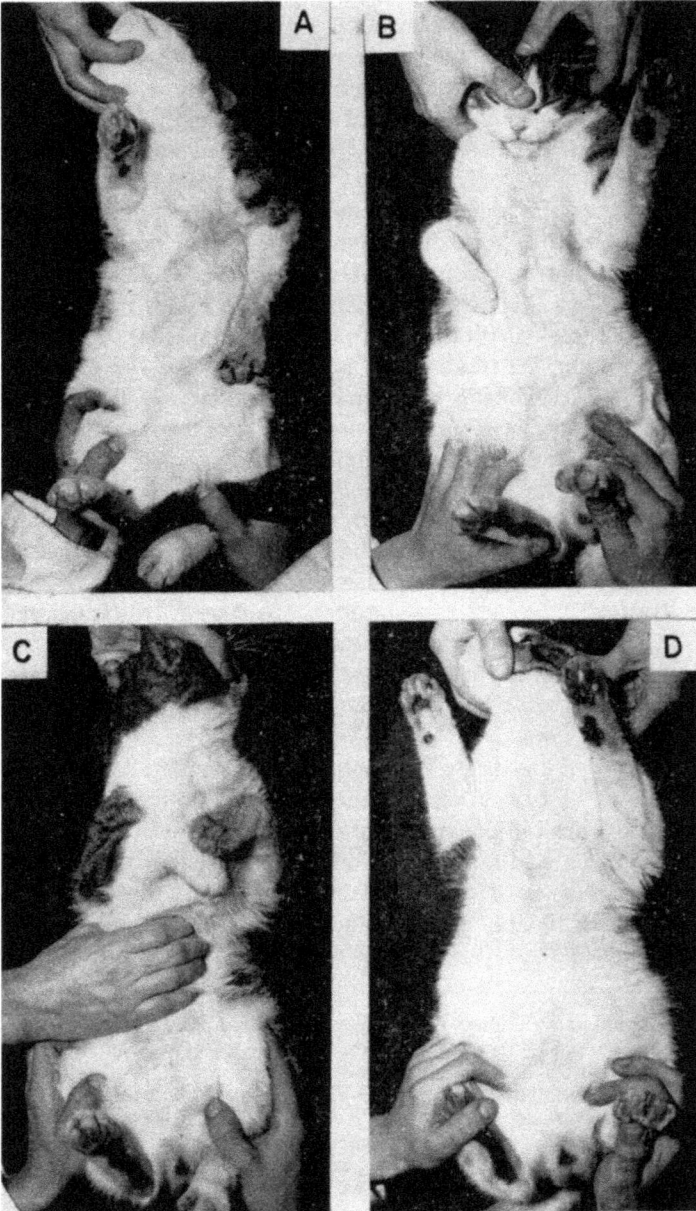

tomy and immobilization of the head on the trunk by means of a plaster cast abolished these reflexes both in the fore and hind limb.

Though not specifically concerned with these reflexes Sherrington (1910), Ranson (1928), Sprong (1929), Pollock and Davis (1930, 1931) and Terzian and Terzuolo (1954) noted an exaggeration of the tonic neck and labyrinthine reflexes in the deafferented limb but failed to observe the reversal of these reflexes. This may be due, in some instances, to the fact that sufficient time was not allowed after section of the dorsal nerve roots for the reversal to develop in the deafferented extremity. Moreover, these investigators studied the reflexes in decerebrate preparations and Teasdall (1950, pp. 213–4, 226) has shown that decerebration performed as long as one year after section of the dorsal roots replaced a well-established reversal by normal, though augmented, labyrinthine and neck reflexes in the deafferented limbs of his animals. Furthermore, these reflexes were usually studied in the deafferented hind limb in which the reversal was much more difficult to elicit and could be clearly demonstrated only when the cats were placed in the vertical plane. This was undoubtedly related to the fact that different postural reflexes oppose each other in the lower extremity but act synergistically in the fore limb (Magnus, 1924, 1926). It is easy to see that such a selective facilitation of reflexes would be of advantage to the animal when landing on its hind limbs in the case of a fall. The mechanism of this facilitation is probably labyrinthine in origin; it has not, however, been analysed in any detail.

How can the reversal of the tonic labyrinthine and neck reflexes which occurs in the deafferented limbs be accounted for? The phenomenon of reversal of responses in denervated structures was made the subject of a series of investigations which will be discussed in chapter IV. It will suffice to say at this time that sensitization by denervation

---

FIG. 26. Reversal of tonic labyrinthine and neck reflexes in a deafferented fore limb of conscious cat studied in supine position (left C 3 to T 3 dorsal roots severed 138 days previously). *A*: with labial cleft 45° above horizontal plane intact (right) fore limb is in "maximal extensor position" but the deafferented (left) fore limb is lying along the side of the body. *B*: flexion of the head causes a reduction in extensor tone of the intact fore limb coincidentally with a maximal extension of the deafferented extremity. *C*: rotation of the head to the left causes flexion of the right (skull) fore limb and a simultaneous flexion of the deafferented (jaw) limb instead of the usual extension of the latter which is shown in Fig. 25. The extensor tone in both fore limbs is greatly reduced in this position. *D*: rotation of the head to the right reverses the position of the fore limbs and the distribution of tone in them. This results in an extension of both fore limbs. (Teasdall and Stavraky, 1955)

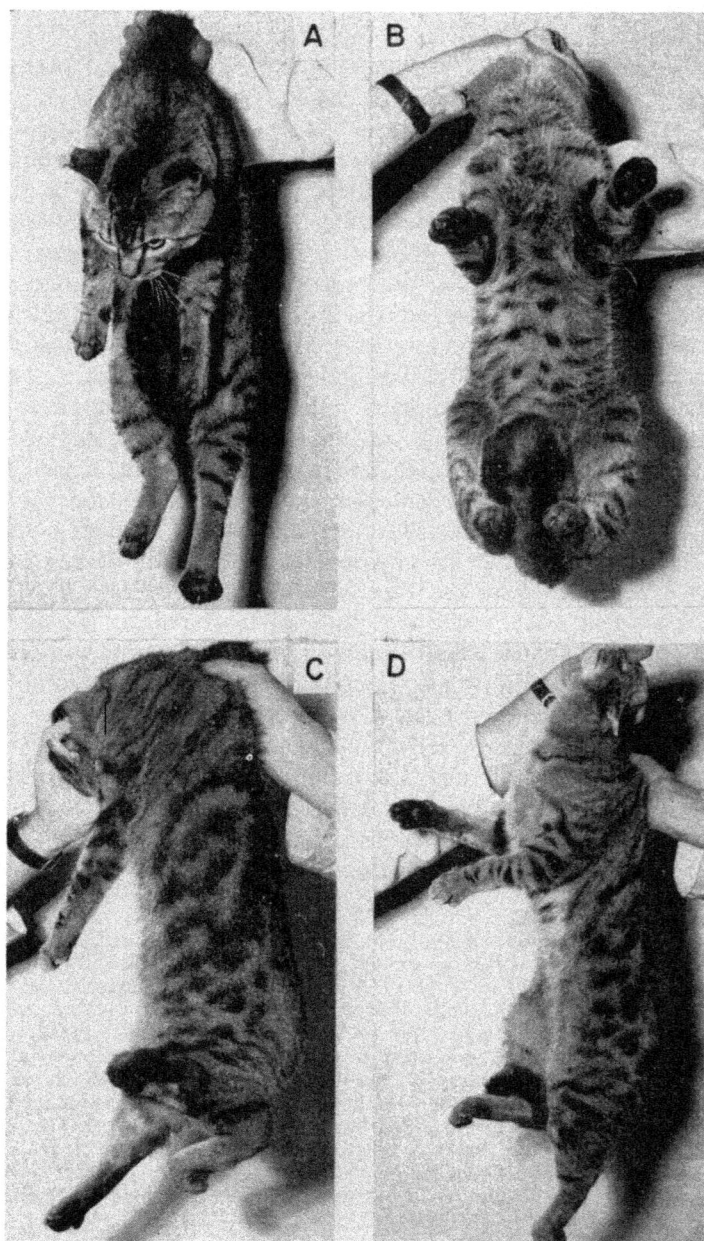

predisposes to reversals which are related in general to an increment of excitation. Following denervation, reversals may be demonstrated more readily both in the peripheral structures and in the central nervous system and apparently are an integral part of the state of supersensitivity.

This interpretation can account for many of the observed facts. Thus when the cat was placed in the supine position, the labyrinthine proprioceptors were maximally stimulated. The resultant excitation of the intact spinal neurones by impinging vestibulo-spinal impulses caused extension of the corresponding limb, while a similar excitation of neurones sensitized by a preceding section of the dorsal roots was excessive and led to an inhibition of the supersensitive neuronal pool—the extensor tone of the deafferented limb becoming reduced and flexion supervening. If the animal was placed in a prone position, stimulation of the labyrinthine proprioceptors was minimal: the vestibulo-spinal impulses thus induced were insufficient to cause a marked discharge of the spinal neurones innervating the extensor muscles of the intact fore limb, and accordingly that limb remained in a semiflexed position. On the other hand, because of the sensitization of the deafferented neurones on the opposite side of the cord, these neurones were excited by the weak vestibulo-spinal volley, and so the deafferented fore limb was extended. Reversal of the tonic labyrinthine and neck reflexes in a chronically deafferented limb may thus be attributed to sensitization of the deafferented spinal neurones to vestibulo-spinal impulses. Whether it was due exclusively to a sensitization at the spinal level or whether the deafferentation had also affected some higher mechanisms is a matter for future studies. It is interesting that in an investigation of children with amaurotic familial idiocy de Kleijn (1923) described a disturbance of tonic labyrinthine and neck reflexes which may have been related to that seen by Teasdall (1950) and Teasdall and Stavraky (1955).

4. DEAFFERENTATION OF SOME BULBAR STRUCTURES AND SENSITIZATION

An early recognition of the fact that deafferentation causes a sensitization of the corresponding brain stem structures as well as of spinal motor

FIG. 27. Tonic labyrinthine and neck reflexes in intact and deafferented hind limb of the conscious cat in the vertical plane. *A* and *B*: before section of the dorsal roots, flexion of the head results in extension of both hind limbs, and extension of the head leads to a flexion of the hind limbs with corresponding changes in tone. *C* and *D*: show reversal of these reflexes in the left hind limb 373 days after intradural section of the L 2 to S 3 dorsal roots. (Teasdall and Stavraky, 1955)

neurones was made by Spiegel and Démétriades (1925) in their analysis of Bechterew's nystagmus. After unilateral labyrinthectomy or after severance of the VIIIth nerve, the nystagmus caused by these procedures gradually subsides. Bechterew (1883) noted that an extirpation of the remaining labyrinth at this time resulted in a nystagmus towards the side of the first labyrinthectomy as though this labyrinth were still present. Magnus and de Kleijn (1923) confirmed this observation and found that the nystagmus persisted after removal of the forebrain. Spiegel and Démétriades (1925) showed that destruction of the diencephalon, the roof of the midbrain, the cerebellar, and the vestibular nuclei on the side of the second labyrinthectomy did not prevent the occurrence of this nystagmus, but it disappeared after the destruction of the vestibular nuclei on the side of the first labyrinthectomy. They concluded that the vestibular nuclei on the side of the chronic labyrinthectomy became sensitized by the operation thus compensating for the loss of the labyrinth. In a recent electrophysiological analysis of the compensation which takes place after unilateral vestibular ablation, Van Eyck (1956) formulated a concept of the mechanism by which this may be achieved. He emphasized the dependence of this phenomenon on reciprocal innervation and concluded that a sensitization of the vestibular nuclei is responsible for it.

Support for this concept was provided in a study by Cook and Stavraky (1952). They showed that sensitization of the vestibular nuclei and possibly of other brain-stem formations may also be produced by partial cerebellar ablation. Various regions of the cerebellum were removed in cats, and after their complete recovery the animals were convulsed by means of intravenous injections of pentylenetetrazol or acetylcholine. Removal of half the cerebellum, with a histologically verified extension of the lesions into the ipsilateral bulbar nuclei, caused the cats to lean to the side of the ablation, and sometimes to lie on, or roll to, that side. Compensation progressed, until, one year after the operation, tremor and ipsilateral ataxia were the only outstanding residual signs. At that time injections of pentylenetetrazol or of acetylcholine reproduced, during the convulsions, the original syndrome of cerebellar removal, the animal falling, rolling, staggering and circling to the side of the operation (Figs. 28, 29 A).

After ablation of the cerebellar nuclei with and without decortication of half the cerebellum but with no extension of the lesions into the bulbar nuclei, the cats were ataxic on the side of the lesion but leaned and staggered to the opposite side for one to three weeks. After complete recovery from the leaning and staggering, pentylenetetrazol and

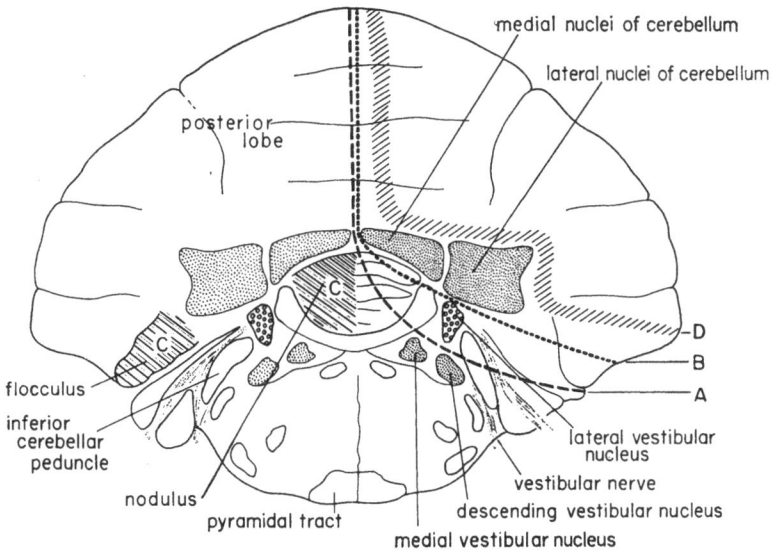

FIG. 28. Cross section of the cerebellum and brain stem of the cat showing schematically the extent of the lesions described in the text. *A*: semidecerebellation with involvement of the vestibular nucleus. *B*: semidecerebellation or lesion in cerebellar nuclei without involvement of vestibular nuclei. *C*: ablation of the flocculo-nodular lobe. *D*: decortication of corpus cerebelli. (Cook and Stavraky, 1952)

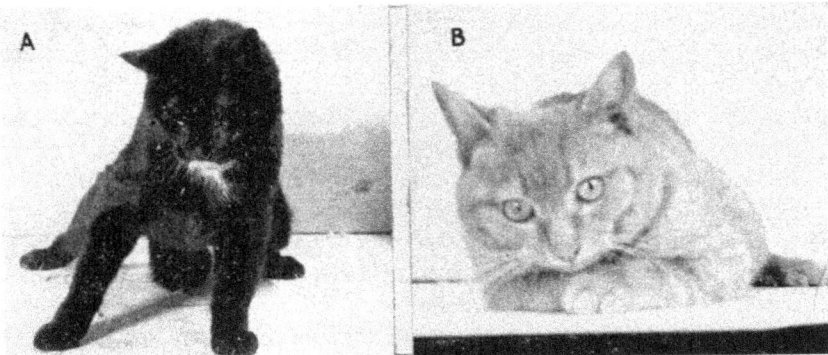

FIG. 29. Typical attitude of semidecerebellate cats with and without extension of the lesion to the bulbar nuclei. *A*: left semidecerebellation of two and a half months' standing with a degenerative change in the left vestibular nuclei. *B*: left semidecerebellation of one week's standing with complete extirpation of the cerebellar nuclei but without any change in the cells of the vestibular nuclei. In both animals the lesions were histologically verified. (Cook and Stavraky, 1952)

acetylcholine caused convulsions in these animals during which they fell, rolled, staggered and circled to the side opposite the cerebellar lesion (Figs. 28, 29 B). Ablation of the flocculo-nodular lobe (partial or complete) gave results essentially similar to those seen after the ablation of the cerebellar nuclei. On the other hand, various cortical lesions of the corpus cerebelli did not influence the direction of the falling and rolling of the animals during convulsions.

The forced movements of the cats to the same side as the lesion following a semicerebellectomy in which the ipsilateral bulbar structures were damaged were interpreted as due to the activity of the uninjured and, therefore, relatively predominant contralateral vestibular nuclei, and possibly other brain-stem nuclei. The deviation of the animals to the side opposite the ablation following unilateral lesions of the cerebellar nuclei or of the flocculo-nodular lobe in which no detectable damage to the bulbar structures was incurred may be regarded as resulting from a partial isolation and consequent sensitization of the ipsilateral vestibular nuclei and other bulbar formations.

Two other instances of sensitization of sensory nuclei following de-afferentation have been described by Chavez and Spiegel (1957). Recording bilaterally the electrical activity of the substantia gelatinosa trigemini from the exposed floor of the fourth ventricle in the cat, they found that section of a fifth cranial nerve centrally to the Gasserian ganglion resulted in a marked reduction of the potentials in the homo-lateral nucleus. However, with extension of the time interval between section of the nerve and the recording (up to forty-five days), the electrical activity recovered to such an extent that in some cases the discharges of the deafferented nucleus and those of the control nucleus could hardly be distinguished in spite of the fact that the recordings were done under nembutal anaesthesia which is known to depress, preferentially, denervated structures.

In another set of experiments, electrodes were implanted into the external geniculate body and the spontaneous electrical activity was studied in chronic experiments lasting up to forty-six days after bilateral section of the optic nerves. Elimination of retinal impulses resulted in a decrease in the spontaneous fluctuations of electrical potential, but there were also signs of hyperactivity, such as the appearance of fast waves of higher voltage than before the operation. Although the lateral genicu-late body is a diencephalic structure, these results were comparable to the ones obtained in other sensory nuclei.

From these experiments the conclusion was drawn that hypersensi-tivity of the cells of the geniculate body following the interruption of visual paths will induce only slight excitatory phenomena in the central optic system of man which could occur only for short periods of time because of the supervening transneuronal degeneration (see page 91). Of greater importance to human pathology is the fact that the neurones of the trigeminal system were able to recover from the initial depression induced by the section of the fifth nerve. According to Chavez and Spiegel (1957): "In most cases of tic douloureux elimination of peripheral trigeminal impulses seems to reduce this bombardment of the respective parts of the diencephalon sufficiently in order to abolish or at least to reduce the pain. However, our finding that the cells of the second neuron i.e. the cells of origin of the quinto-thalamic pathways may reach a normal level of discharges some time after elimination of the impulses from the peripheral neuron may explain that retrogasserian section of the trigeminal root fails in some instances of tic douloureux to relieve the attacks."

5. THALAMIC LESIONS AND CORTICAL SUPERSENSITIVITY

The ability of a partially isolated cerebral cortex to carry on spontaneous rhythmic activity and to develop bursts of electrical discharges was shown by Spiegel (1937), Bremer (1938), Swank (1949), Kristiansen and Courtois (1949), Burns (1951), Henry and Scoville (1952), Echlin *et al.* (1952), and others. Definite proof that isolation of the sensory cortex leads to an enhancement of its sensitivity was furnished by Spiegel and Szekely (1955). The method which they employed was complementary to that used by Miller, Stavraky and Woonton (1940) who showed that local application of acetylcholine to the eserinized sigmoid gyri of the cat causes a synchronization and exaggeration of the electrical activity of the appropriate areas of the cerebral cortex and the appearance of strychnine-like spikes in the electrocorticogram. Spiegel and Szekely (1955) implanted electrodes epidurally over the posterior sigmoid gyri of the cat and placed an electrolytic lesion in one posteroventral thalamic nucleus involving skin, muscle and taste afferents. The electrical activity of the cortex was studied before and after the operation. Immediately following the thalamic lesion there was a depression of cortical activity over the ipsilateral posterior sigmoid gyrus. After three to six days, however, the cerebral cortex on that side became hyperexcitable, judged by its reaction to intramuscular injections of methacholine and to intravenous injections of pentylenetetrazol and of bulbocapnine (Fig. 30). Greater cortical activity was also present in the "resting state": the potential fluctuations from the posterior sigmoid

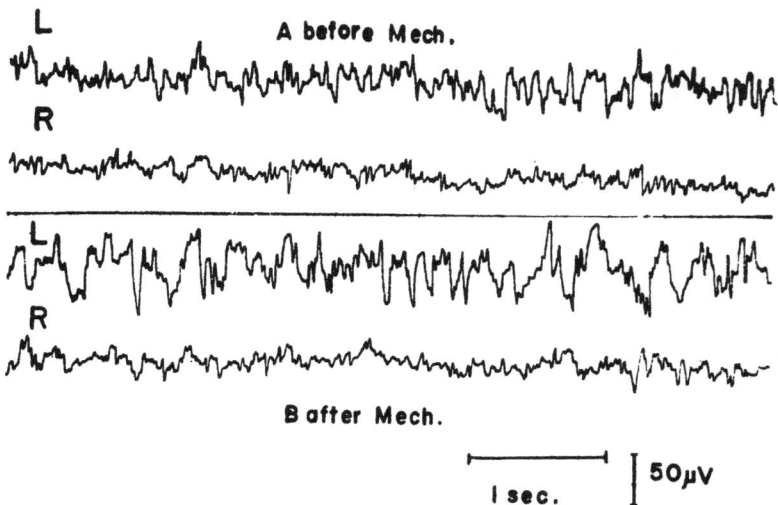

FIG. 30. Electrocorticogram of left (L) and right (R) posterior sigmoid gyri of cat in bulbocapnine catalepsy taken six days after the destruction of the left posteroventral thalamic nucleus. Note dominance of the electrical activity on the side of the lesion (*A*) before methacholine (Mech.), and a further selective potentiation of the activity on that side after the injection of 1 mgm. of methacholine (*B*). (Spiegel and Szekely, 1955)

gyrus on the side of the lesion were often found to be of higher amplitude and greater frequency than those led off from the opposite side. The cortical hyperactivity on the same side as the thalamic lesion became even more apparent on stimulation of peripheral receptors, for instance, by immersing the hind leg of the animal into hot water. Commenting on these results Spiegel and Szekely (1955) pointed out that development of supersensitivity in sensory cortical areas after thalamic lesions may have a bearing on the development of pain and hyperpathia which occur in cerebral lesions extending into the white matter below the sensory cortex.

6.   CONCLUSIONS AND CLINICAL IMPLICATIONS

From the evidence presented the general conclusion can be drawn that interruption of afferent connections results in two distinct disturbances. The immediate effect is a lack of facilitation which is brought about by the disconnection of the peripheral receptors from their central attachments. It is dependent upon an anatomical defect and in the absence of regeneration of nerve fibres is not subject to further readjustment or change. The second effect is an exaggeration of the sensitivity of the denervated neurones. It is this central effect which is subject to progressive readjustments and changes. In time it may compensate for the decreased facilitation from the periphery and lead to an overaction of the partially isolated regions. The effects which the lack of facilitation and changes in sensitivity of the deafferented neurones exert on tone have been described. However, other clinically important features of the activity of the central nervous system may be influenced by the ensuing readjustments. Two of them will be considered.

A. *An interpretation of sensory ataxia*

Sensory ataxia may be broadly defined as loss of precision of voluntary movements, or, more specifically, as an "inability to regulate the direction, rate, force and extent of these movements" (Ruch, 1955, p. 341).

Whether sensory ataxia results from the involvement of the posterior nerve roots or of the dorsal columns of the spinal cord is a matter of conjecture. The former is often favoured, but involvement at either site can, apparently, provoke some degree of the disorder. Physiologically, ataxia is most commonly attributed to the loss of proprioceptive sensibility which, through the muscle spindles and Golgi tendon organs, regulates the contractions of the muscles, giving them smoothness and precision and which, through the sense of position of the limbs, controls the extent and direction of the desired voluntary movement. Sensory ataxia is thus related to the study of the effects of deafferentation.

The great advances which have been made in recent years in our knowledge of the functions of the muscle and tendon proprioceptors overshadowed the observations on changes of sensitivity of the spinal motor neurones which have been summarized in this chapter. The role of the gamma-efferent system, the inhibitory and facilitatory influence of the reticular formation on this system and the part played by them in the regulation of the activity of muscle spindles have been elucidated in great detail. These findings are pertinent to the problem of sensory ataxia because they are intimately related to the mechanism of regulation of movement and, without this knowledge, no complete concept of a disturbance of movement can be worked out. However, it may be recalled that Liljestrand and Magnus (1919) abolished proprioceptive sensibility of different muscles by the injection of appropriate doses of procaine into the triceps, biceps, supraspinatus, infraspinatus, subscapularis and shoulder, elbow and wrist joints of cats; but they failed to produce ataxia. The movements were studied on the background of voluntary walking and standing which were retained in spite of some loss of tone and of the tendon jerks. A characteristic disturbance of motor activity was noted, but it was distinct from an ataxia induced by chronic deafferentation. Bremer (1928) repeated these experiments in the hind limb of the cat and came to similar conclusions, and Walshe (quoted by Bremer, 1928) failed to induce ataxia by injecting procaine into the muscles of human subjects. The removal of proprioceptive sensibility was confirmed in Bremer's experiments by the fact that procaine abolished decerebrate rigidity without affecting the performance of the motor neurones as judged by the study of the crossed extensor and of the vestibular reflexes induced by linear acceleration.

More recently, it was shown in man by Libet, Feinstein and Wright (1955) that local anaesthesia of a tendon removes autogenetic inhibition through which the power of isometrically recorded voluntary contractions of the anterior tibial muscle in certain positions of the limb was augmented. Local anaesthesia thus removes autogenetic inhibition but fails to produce typical ataxia.

Drake and Stavraky (1948 b) have demonstrated that after deafferentation spinal neurones become supersensitive to chemical agents. This supersensitivity develops both to stimulating agents and to barbiturates, other anaesthetics, cold and various adverse conditions. An exaggerated depression cannot be ascribed to the removal of autogenetic inhibition. Similarly in the experiment of Luco and Eyzaguirre (1952) the dorsal and ventral roots were cut on two sides simultaneously, yet a greater sensitivity of the motor neurones to strychnine developed on the side on which the dorsal root was sectioned proximally to the ganglion. This emphasizes the fact that the degeneration of afferent connections, and not the sensory input, determines the sensitivity of the motor neurones. Furthermore, the first effect of deafferentation is a depression of spinal neurones; the exaggerated responses appear after a considerable lapse of time and the excitability of these neurones and the muscular tone increases progressively over a period of weeks and months.

When the stage of augmented excitability reaches its maximum, not only an exaggeration but a profound derangement of postural reflexes becomes apparent, an example of the latter being a reversal of the tonic labyrinthine and neck reflexes (Teasdall and Stavraky, 1955). Both an exaggeration and a reversal of the effects of stimulation of the cortico-spinal tracts may also be seen (Teasdall and Stavraky, 1953). These phenomena are probably important contributory factors in the pathogenesis of ataxia.

Sensory ataxia is, therefore, not due to deficiency phenomena alone, such as lack of sensory perception, hypotonia or absence of autogenetic inhibition, but to a composite functional alteration in which lack of sensibility and a change in the excitability of the deafferented neurones together lead to an instability of responses of the spinal centres to voluntary impulses reaching them through the cortico-spinal tracts. This instability is determined by: (1) a disbalance of reflex postural adjustments—absence of some (such as, muscle spindle and tendon organ regulation) and exaggeration or reversal of others (labyrinthine and neck reflexes)—on the background of which voluntary movements are brought about; (2) by an "overshooting" of the limbs as a result of the exaggerated responses of supersensitive neurones to a given train of voluntary impulses; and (3) by the lack of the sense of position which makes orientation as to the result of the movement impossible.

B. *Causalgia, thalamic syndrome and hyperpathia of subcortical lesions*

As summed up by Grinker and Bucy (1949, p. 383), "while paresthesias or dysesthesias are commonly produced by neurological lesions of diverse segments of the peripheral and central nervous system, their explanation is as yet impossible. They may represent irritative, or release phenomena, or both, modified by cortical associations." If one analyses the various trends of thought devoted to the problem of "central pain," the point that comes up most constantly in the different concepts of its origin is the duality of the systems of sensibilities which are affected. Dissociations of sensation in which some pain is selectively preserved are most commonly found to result in the qualitative change of a heightened disagreeableness of the sensation (Stein *et al.*, 1941 a, b; Lewis, 1942; Bremer, 1954; and others). The concept of dissociation of sensation implies a "partial deafferentation" and the latter, as shown in this section, leads to supersensitivity of central mechanisms upon which the sensory fibres project. Thus, it may, in turn, be related to the phenomenon of sensitization by denervation and may be discussed in these terms. The plausibility of such a concept is enhanced by the findings of Kennard (1953) who showed that destruction of some sensory neurones in the spinal cord of the cat potentiates the activity of the remaining units, this resulting in hyperalgesia.

The observations of Head *et al.* (1905) on prothopatic and epicritic sensibility as expanded in physiological and anatomical investigations (Erlanger and Gasser, 1930; Gasser, 1934, 1937; Lewis and Pochin, 1937, 1938; Ranson and Billingsley, 1916; Ranson and Davenport, 1931; and others) led Kendall (1939) to the assumption that sensory impulses in a peripheral nerve are transmitted by two independent pathways distinguishable by fibre size and rate of conduction. He suggested that "central pain" is the result of interruption of fast conducting fibres in the spinal cord and brain stem by lesions which leave intact the slow conducting fibres. According to this view, the delayed disagreeable sensation normally conducted by the slow fibres is blocked by the arrival at the thalamus of impulses which travel by faster paths. If the fast fibres are destroyed, the inhibitory process fails, and the diffuse, poorly localizable pain is felt. Experimental substantiation of this view came from Landau and Bishop (1953) and Bishop and Landau (1958) who, working with asphyxial nerve block, concluded that slow pain impulses conducted along the C fibres give rise to dull, burning, disagreeable pain when not preceded by the delta fibre pain impulses.

Causalgia, which is characterized by pain following peripheral nerve injury in the absence of marked sensory impairment, was interpreted by Kendall (1939) as being dependent on factors similar to those concerned in the production of pain in central lesions. The finding that sympathectomy or sympathetic nerve block is beneficial in causalgia (Leriche, 1939; Livingston, 1947; Mayfield, 1951; Ash, 1952) does not contradict the general hypothesis based on the interrelation of different systems of sensory nerve fibres but rather adds another component to it.

The wide use of surgical section of spino-thalamic tracts in patients with intractable pain led to some pertinent observations. Drake and McKenzie (1953) found that when high division of the pain fibres (mesencephalic tractotomy) was carried out as suggested by Walker (1942 a, b), there was complete relief from pain early in the post-operative period, but within three days to two weeks a new response developed to pin prick, deep pressure or thermal stimulation. Upon such stimulation the patients complained of agonizing, poorly localized pain. Half of the patients also developed spontaneous burning pain, which was in some instances worst in the region of the original pain. Drake and McKenzie (1953) pointed out that in accordance with the view held by Lindsley, Schreiner, Knowles and Magoun (1950) and Starzl, Taylor and Magoun (1951) the afferent systems give off collaterals into the reticular formation as they pass up the brain stem and concluded that a conventional chordotomy interrupts the afferent fibres below the origin of the collaterals, whereas mesencephalic tractotomy spares them and thus provides a pathway for the development of post-operative dysaesthesia. Its gradual onset was not explained.

The most essential feature of the thalamic syndrome of Déjerine-Roussy is the disorder of sensation which is, again, a partial anaesthesia on the corresponding side of the body and paroxysms of so-called spontaneous pain. This latter is not an exclusive feature of the thalamic syndrome; as mentioned before, similar disorders of sensation may be encountered in lesions of the brain stem, spinal cord and even of peripheral nerves. Furthermore, according to Weir Mitchell (1897), who apparently was the first to describe the thalamic syndrome, pain felt in this instance is not truly spontaneous but is brought about by movements of air and other extraneous stimuli, and this observation finds support in the researches of Spiegel and Wycis (1953) and Spiegel *et al.* (1954) who came to the conclusion that bombardment of the diencephalon by afferent impulses is an essential factor in thalamic pain. Head and Holmes (1911), impressed by the hyperemotional reactions of the afflicted side, attributed them, as well as the pain in thalamic lesions, to a loss of cortico-thalamic inhibition. Kendall (1939) felt that both spino-thalamic and cortico-thalamic inhibitory processes may play their part in thalamic pain, but Spiegel and Szekely (1955) point out that the sensory cortex can play only an accessory role in this condition since its undercutting is unable to eliminate the pain of the thalamic syndrome. However, these investigators believe that the hyperpathia described in cerebral lesions by Davison and Schick (1935) may be due to a sensitization of cortical areas produced by the extension of the lesion in the white matter below the sensory cortex thus interrupting afferent paths from the thalamus. That isolation of an area of the parietal lobe by destruction of adjacent regions may produce hyperpathia was shown by Peele (1944). Also noteworthy is the effect of hallucinogens: mescaline is reported to affect phantom limbs in human amputees much earlier and more severely than the rest of the body (Zador, 1930: Mayer-Gross, 1951). An exception to this pattern were the visual hallucinations which decreased in frequency and vividness after removal of an eye. These findings were explained on a functional basis, but fit in well with the concept of supersensitivity due to partial denervation. The eye presents a special case in view of the fact that section of the optic nerve in monkey and in man leads to a total isolation of neurones in the lateral geniculate body which results in severe atrophic changes in the nerve cells (Glees and Clark, 1941; Cook, Walker and Barr, 1951) and in an over-all reduction of the electrical activity of the latter (Chavez and Spiegel, 1957) in spite of the presence of some signs of hyperactivity.

In summary, paraesthesias, dysaesthesias and hyperpathias of central origin, whether they are of the type of causalgia or of thalamic pain or whether they are due to mesencephalic tractotomy or to cortical lesions, seem to be associated with partial denervation of certain regions of the central nervous system. There is ample evidence that deafferentation renders these regions "supersensitive" to chemical stimulating agents and to nerve impulses reaching them by way of the remaining connections. Such a supersensitive state would account well for the paroxysms of bizarre, poorly localized, but exaggerated and therefore intolerable, pain that is felt when these areas of the central nervous system are activated through their remaining connections. The effect of sensitization is shown particularly clearly in mesencephalic tractotomy (Drake and McKenzie, 1953) which abolishes the original pain for which it is done but leads to a gradual development of an insufferable dysaesthesia.

| Reversals of the Effects of Nervous

and Chemical Stimulation

The great lability of the responses of denervated structures compared with their intact counterparts stood out from the studies on supersensitivity. Partially isolated regions of the central nervous system became not only more sensitive to stimulating and depressing chemical agents and to nerve impulses reaching them by way of the remaining connections, but often were depressed to a far greater extent than were the intact areas of the brain or spinal cord by cold, fatigue, operative procedures and anaesthesia and by such ˙diseases as distemper. Any one of these factors often led to a reversal of the sensitivity, the control side responding more readily and more markedly to various types of stimulation than the previously denervated side (Stavraky, 1947; Drake and Stavraky, 1948 b). The same results were obtained when the convulsibility was compared in intact and operated animals (Stavraky, 1943, 1947; Teasdall and Stavraky, 1950; Teasdall, 1950). Furthermore, at the peak of sensitivity of the isolated neurones a reversal from excitation to inhibition was seen to take place (Teasdall and Stavraky, 1953, 1955). Thus, application of excessively strong stimuli or the superposition of a stimulus on top of a pre-existing excitatory state resulted in the inhibition of tone or in abrupt cessation of movements. Further study of the mechanism of this reversal was therefore indicated.

1. REVERSALS AND THE AUTONOMIC NERVOUS SYSTEM

A. *Reversals as observed in the submaxillary salivary gland*
   One of the first instances of a reversal in the effect of stimulation of an autonomic nerve was described by Fröhlich and Loewi (1906, 1908). They reported that in decerebrate cats the vascular response of the submaxillary salivary gland to stimulation of the chorda tympani was

altered by the administration of nitrites, the chorda tympani under these conditions causing diminution in the rate of blood flow through the gland instead of the usual vasodilatation. Pilocarpine sensitized the gland to this reversal of the vascular response. Sympathectomy and nicotine in amounts which paralysed the ganglia did not affect it, but atropine abolished the response. Fröhlich and Loewi concluded that the phenomenon was due to the existence of vasoconstrictor fibres in the chorda tympani, and that the action of these fibres became evident when the effect of the dilator nerves was eliminated by nitrite poisoning.

Bayliss (1908) was unable to confirm these findings, but, the experiment was repeated successfully by Dale (1930). Amyl nitrite was not unique in this respect, and histamine and low blood pressure also predisposed to an apparent reversal of the blood flow through the submaxillary gland on stimulation of the chorda tympani. Dale (1930) concluded, however, like Gesell (1919, 1920), that the effect could be caused by an escape of fluid into the secretion and not by an actual vasoconstriction.

Other instances of reversal pertaining to the submaxillary gland became known. MacKay (1927) reported that histamine, which by itself causes vasodilatation, when injected after pilocarpine diminished the blood flow through the gland and at the same time inhibited the pilocarpine-induced secretion. Anochin and Anochina-Ivanova (1929) found that in dogs small quantities of acetylcholine accelerated pilocarpine secretion, and that large doses of this agent caused inhibition of the secretion induced by pilocarpine. An analogous relation was found by Stavraky (1942) to prevail with adrenaline, which in small doses augmented the secretion and vasodilatation caused by pilocarpine, eserine, or methacholine, but in large quantities caused vasoconstriction and a reduction of secretion.

Furthermore it was found by Stavraky (1934) and Graham and Stavraky (1953 a) that, after inhalation of amyl nitrite or injection of large quantities of eserine, chorda tympani stimulation often caused a simultaneous diminution of secretion and blood flow through the gland. Thus Gesell's and Dale's hypothesis that an apparent diminution of the blood flow took place due to the escape of fluid into the secretion became untenable.

A reversal of the effect of chorda tympani stimulation by eserine was of particular interest in view of the anticholinesterase activity of the latter. It was established that acetylcholine is liberated at the neuro-effector junctions on stimulation of the chorda tympani (Babkin, Alley and Stavraky, 1932; Babkin, Gibbs and Wolff, 1932; Emmelin and

Muren, 1950 a; Henderson and Roepke, 1933 a, b) and the possibility arose that the reversal was related to the effect of acetylcholine acting in excess. This was demonstrated to be the case by Graham and Stavraky (1953 a) on close-arterial injections of acetylcholine (Fig. 31). The effect of acetylcholine was reversed not only when it was injected in excessive amounts, but preceding injections of acetylcholine reversed the effect of stimulation of the chorda tympani. Eserine reversed the effect of acetylcholine in the same way as it did that of stimulation of the chorda tympani. An exhaustive comparative study of the action of a large group of anticholinesterases on the submaxillary gland undertaken by Dirnhuber and Evans (1954) yields strong support to these findings. Similarly, Graham and Stavraky (1953 a) found that small quantities of adrenaline (10^{-8} — 10^{-2} µg.) injected intra-arterially caused vasodilatation in the submaxillary gland, whereas relatively large quantities of this agent (10^{-1} — 5 µg.) caused vasoconstriction and secretion. Stimulation of the cervical sympathetic trunk can also yield a dual effect. It has been known since the early investigations of the English school that stimulation of the sympathetic nerve causes vasodilatation and secretion in the submaxillary gland of the cat under ether and chloroform anaesthesia. However, under light pentobarbital anaesthesia the same stimulation usually results in transient vasodilatation, which is followed by vasoconstriction if the stimulation is continued for any length of time.

B. *Denervation and reversals in the submaxillary salivary gland*

On the whole the observations on an acutely isolated gland favour the conclusion that the autonomic nerves and their chemical mediators may produce dual effects, the reversal of the initial response being brought about by excessive stimulation. How would chronic denervation of the gland affect its responses? Fleming and MacIntosh (1935) reported that, in the cat after section and degeneration of the parasympathetic nerve supply, acetylcholine injected intravenously in doses of 500 µg. evoked a greater secretion of saliva from the control submaxillary gland than from the denervated one. Comparable observations were reported by Pierce and Gregersen (1937) for the dog on the intravenous injection of 100 mgm. of acetylcholine. These observations constituted an apparent exception to the "law of denervation" formulated by Cannon (1939) and have not been satisfactorily accounted for, in spite of the fact that Wills (1942) reported a lower threshold and a greater secretion of saliva after chronic denervation of the submaxillary gland in the cat after intra-arterial injection of small quantities (0.03 — 2.0 µg.) of

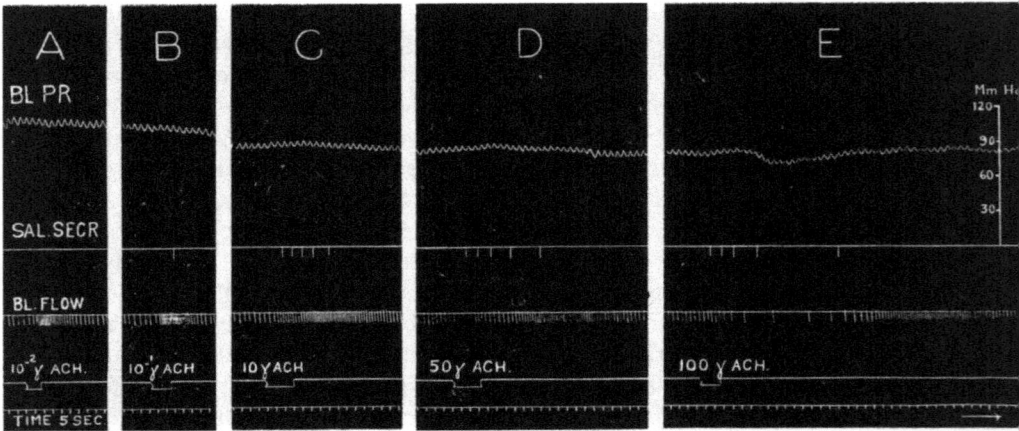

FIG. 31. Different responses of the submaxillary salivary gland of the cat evoked by close arterial injections of increasing quantities of acetylcholine (ACH.) under chloralose-urethane anaesthesia. Secretion (Sal. Secr.) and blood flow (Bl. Flow) recorded in drops of equal volume; blood pressure (Bl. Pr.) recorded from the carotid artery on the opposite side. Note a change from vasodilatation (*A* and *B*) to vasoconstriction (*E*) with increasing quantities of acetylcholine. (Graham and Stavraky, 1953)

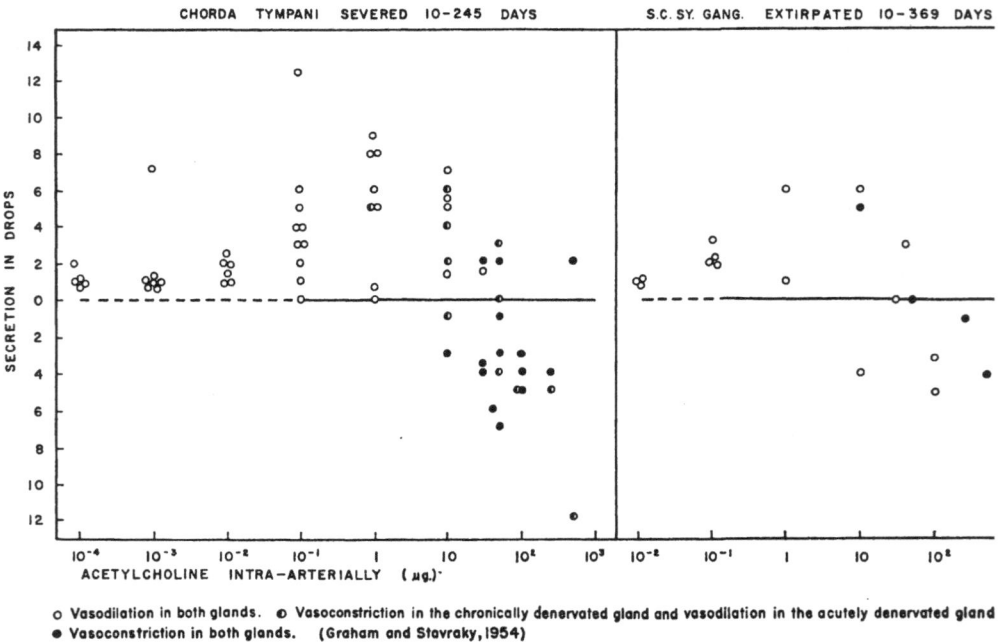

o Vasodilation in both glands. ⊙ Vasoconstriction in the chronically denervated gland and vasodilation in the acutely denervated gland ● Vasoconstriction in both glands. (Graham and Stavraky, 1954)

FIG. 32. Vasomotor and secretory effects of different quantities of acetylcholine (close arterial injections) as observed in the chronically and acutely denervated submaxillary gland of the cat. The difference between the volume of secretion of the chronically denervated gland and that of the acutely denervated gland is plotted relative to the secretion of the acutely denervated gland which is represented by the horizontal "0" line. The point of change over from an interrupted line to a solid one indicates the secretory threshold of the acutely denervated gland.

acetylcholine. His findings were corroborated by Emmelin and Muren (1951) with intravenous administration of acetylcholine in doses ranging from 1 to 10 μg./kgm. Furthermore, chronic denervation, parasympathetic or sympathetic, has been shown to increase the secretory response of the submaxillary gland to sympathomimetic agents (Fleming and MacIntosh, 1935; Simeone and Maes, 1939; Emmelin and Muren, 1950 b, 1951 a, b, 1952; Emmelin, Jacobson and Muren, 1951; Emmelin and Stromblad, 1951; Emmelin, 1952 a, b). Graham and Stavraky (1954 c) reinvestigated the relation of the reversal in the effects of nervous and chemical stimulation of the gland to denervation, combining the technique of close-arterial injections of various agents and the simultaneous recording of blood flow and secretion of saliva. Denervation of one gland, accomplished by aseptic excision of either the chorda tympani or of the superior cervical sympathetic ganglion 10 to 423 days previously, resulted in a greater sensitivity of this gland to the tested drugs than of the control gland, the type of response depending on the degree of stimulation (Fig. 32). The denervated gland had a lower threshold and responded by a greater volume of secretion and by greater vasodilatation to injections of small quantities of acetylcholine than did the control. On the other hand, large quantities of acetylcholine caused a greater vasoconstriction and reduction of secretion in the denervated gland than in the control.

Similarly, an increased sensitivity of the denervated gland to adrenaline and to noradrenaline was evident from the lowered secretory threshold and the greater volume of secretion when these agents were injected in small or moderate quantities, and from the greater vasoconstriction and more profound reduction of secretion when they were introduced in large amounts. Thus the contradictory findings of Fleming and MacIntosh (1935) and Pierce and Gregersen (1937) on one hand, and those of Wills (1942) and Emmelin and Muren (1951 b) on the other could be accounted for by the different quantities of the stimulating agents used. In special experiments in which a wide range of doses of acetylcholine was injected intravenously this was actually shown to be the case.

The effect of atropine on the acutely and on the chronically denervated gland also varied, depending on the dose of acetylcholine injected or on the quantity of atropine employed. Fleming and MacIntosh (1935) were of the opinion that atropine affected the denervated submaxillary gland less than it did the intact one. Similarly, Altamirano, Fernandez and Luco (1949) described the diminished effectiveness of atropine on the superior cervical sympathetic ganglion and on the

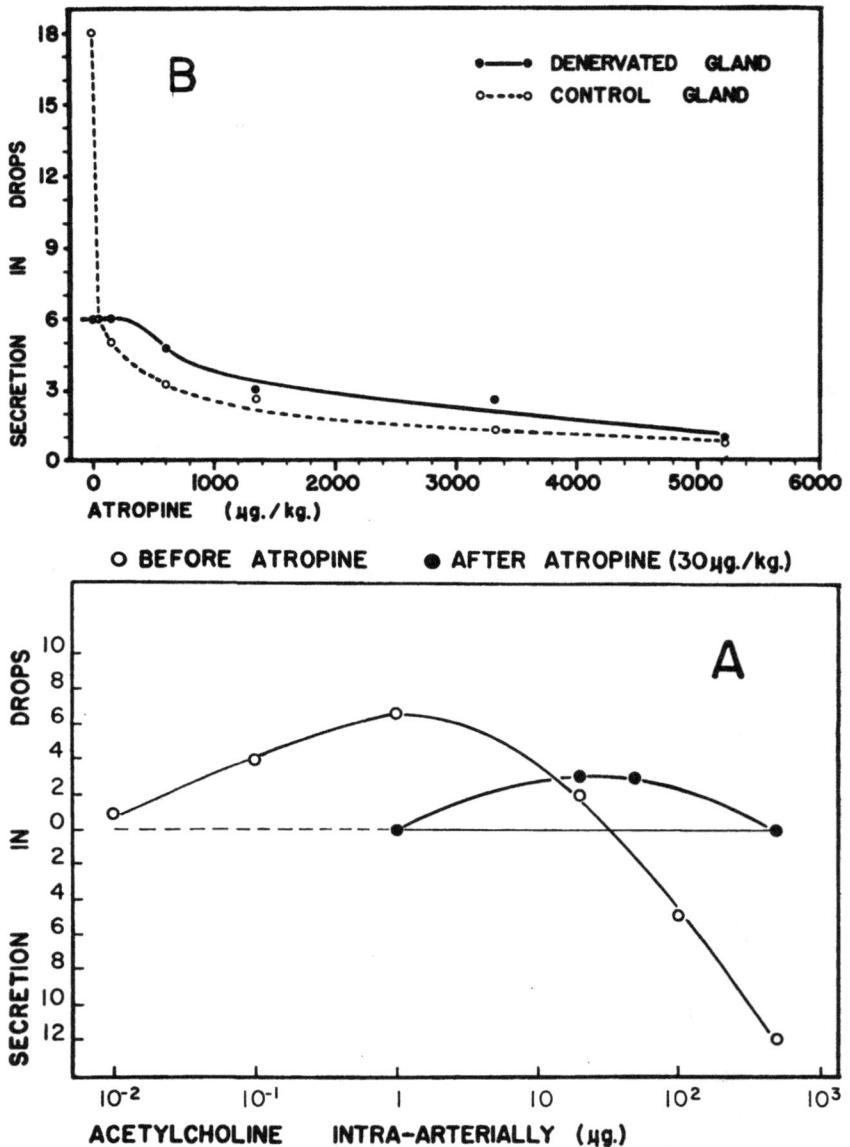

FIG. 33. Effect of atropine on the secretory responses of the submaxillary glands of a cat 245 days after section of the left chorda tympani. *A*: difference between the volume of secretion of chronically and acutely denervated gland plotted as in Fig. 32. Note: the difference between the secretory thresholds of the two glands was abolished and the difference in the volumes of secretion was reduced by the partial atropinization. *B*: effects of repeated close-arterial injections of 500 μgm. of acetylcholine with interposed injections of atropine. Note: before atropine the chronically denervated gland secreted less saliva in response to 500 μgm. of acetylcholine than the control gland. Atropinization reversed the response of the two glands. (Graham and Stavraky, 1954)

nictitating membrane following denervation of these structures. However, when Graham and Stavraky (1954 c) employed a wide range of doses of acetylcholine before and after partial atropinization of the gland, it became apparent that atropine had a profoundly depressing effect on the denervated gland. Minimal doses of atropine were seen to raise the secretory threshold of the denervated gland to acetylcholine without affecting the control (Fig. 33 A). With an intermediate range of doses acetylcholine, after such slight atropinization, produced a greater secretion from the denervated gland, but the difference in the secretory response of the two glands was reduced. Excessive quantities of acetylcholine caused vasoconstriction and diminution of secretion, but the difference in this latter effect was again not as pronounced between the two sides as before atropinization.

A variety of combinations of effects could be seen when the quantities of atropine and acetylcholine were increased, and as shown in Fig. 33 B the effect of very large doses of acetylcholine could be reversed in the two glands by progressive atropinization. These effects of atropine, at least in part, were due to vasomotor responses. The preferential vasoconstriction evoked by large quantities of acetylcholine in the denervated gland was altered by atropinization. This may have created a complicated change in the pattern of the secretory responses of the two glands, in which the secretion of saliva would be greater either from the denervated or from the control gland, depending on how much acetylcholine actually reached each of these two organs.

C. *Reversals in some other organs under autonomic control: their possible mechanism and implications*

In his pioneer studies on the vasomotor effects of acetylcholine, Reid Hunt (1918) observed that weak solutions of acetylcholine augmented the flow from the perfused isolated muscle of the dog and cat, and that strong solutions caused a marked diminution of the flow. Similar results were obtained in the skin of the rabbit abdomen, in a loop of intestine, and in the isolated spleen. These observations were confirmed by numerous investigators in different experimental situations (Fleisch, 1931; Hirose, 1932; Gaddum and Holtz, 1933; Alcock, Berry and Daly, 1935; Sakamoto, 1935; Uchino, 1935; Hattori, 1936; Kohn *et al.*, 1936; Necheles *et al.*, 1936 a, b; Lubsen, 1941 a, b; and others). Gradually evidence has accumulated which indicates that acetylcholine, introduced in small quantities, causes salivary and gastric secretion, acceleration of the heart and dilatation of the coronary arteries, changes in intestinal motility, and contraction of striated muscles. When administered in relatively large amounts, particularly by the intra-arterial route or by local application, acetylcholine causes the opposite effects (Laporta and Saviano, 1943; McDowall, 1946; Peruzzi, 1947, 1948; Prosser, 1940; Sachs, 1937; Spadolini, 1948; Spadolini and Domini, 1940; Welsh, 1948; Katz, Lindner, Weinstein, Abramson and Jochim, 1938; Eckenhoff *et al.*, 1947; Folkow *et al.*, 1948, 1949; Wégria, 1951; Brown, 1937).

Similarly, the activity of the nervous system may be enhanced or depressed by suitable concentrations of acetylcholine (Bonnet and Bremer, 1937 a, b; Sjöstrand, 1937; Kirillova, 1943; Bülbring and Burn, 1941; Feldberg and Vartiainen, 1934; Cannon and Rosenblueth, 1937; Rosenblueth, 1950; and others). A dose-response relation prevails not only for acetylcholine but for adrenaline as well (Raventos, 1935; Clark, 1935; Pucinelli, 1933; Foggie, 1937; Goetz, 1939; Lubsen, 1941 a, b; Burn and Robinson, 1951; Harpuder, Byer and Stein, 1947; Meier and Bein, 1950; and others). Instances of reversal in the effect on various structures of stimulation of autonomic nerves also became known, the coronary arteries attracting special attention (Anrep and Segall, 1926; Katz, Jochim and Bohning, 1938, 1939; Folkow, 1955).

The mechanism of these reversals received much thought. Burn and Vane (1949) and Burn (1950) suggested the possibility that the dual effect of acetylcholine may be associated either with its influence on acetylcholine synthesis in the tissues or with the rate of hydrolysis of acetylcholine in the tissues, the hypothesis being that in the latter instance the blocking of available receptors and the excess of unhydrolysed acetylcholine lead to inhibition. An analogous mechanism supposedly may be involved in the reversal of the action of other substances.

On the other hand, Nachmansohn (1945) and Welsh (1948, 1955) consider that acetylcholine may play an integral part in the energy-producing systems of cells as well as having a trophic influence. In accordance with Welsh's view, acetylcholine may be supposed to reverse its effect, on the basis of its action as a co-enzyme to a "receptive substance" in the cell. Welsh (1948) states that "it is usually possible to demonstrate that acetylcholine in low concentrations has an excitatory action, while at higher concentrations the response is depressed or inhibited. Since the excitatory action on one type of organ or tissue (e.g. smooth muscle of the intestine) may occur over a wide concentration range and be easily recorded or observed, while for another tissue or organ the excitatory range may be narrow and the inhibitory effects prominent (e.g. vertebrate or molluscan heart muscle) it has become customary to think of acetylcholine as having two types of action, ex-

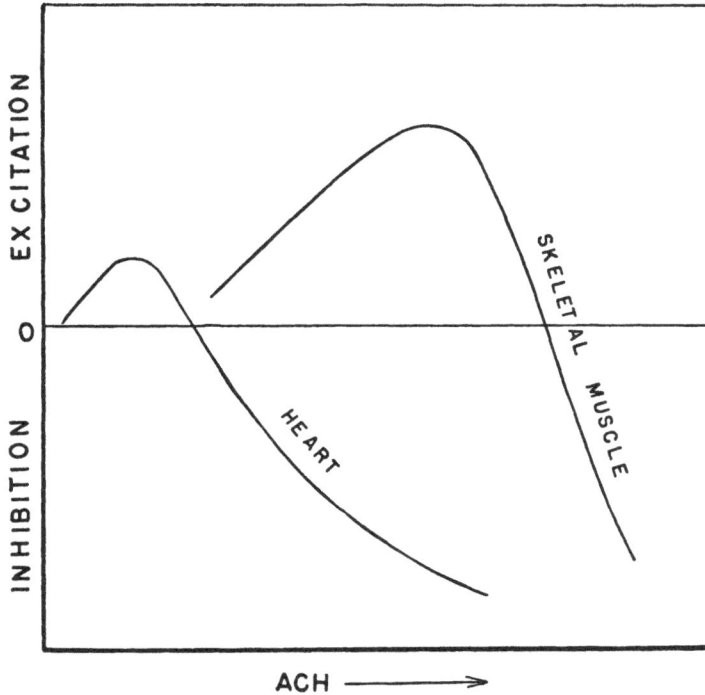

FIG. 34. Scheme showing suggested excitatory and inhibitory effects of increasing quantities of acetylcholine on vertebrate heart and skeletal muscle. (Welsh, 1948)

citatory and inhibitory (or paralytic), separate and distinct from one another." (Fig. 34) Welsh (1955) further points out that the vertebrate heart normally operates with a high level of endogenous acetylcholine which is in the inhibitory range. But that when the residual level of acetylcholine and the rate of synthesis are low, the heart is excited by acetylcholine (Abdon, 1945; Bülbring and Burn, 1949; McDowall, 1946; Spadolini and Domini, 1940; Giachetti, Peruzzi and Spadolini, 1953; and others).

The reverse may prevail in the submaxillary gland. Under the usual experimental conditions the level of endogenous acetylcholine is such that it exerts a stimulating effect which is potentiated by anticholinesterases. This latter effect is attributed to a delayed destruction of acetylcholine in the gland (Beznak and Farkas, 1937; Riker and Wescoe, 1949). However, if the critical level of acetylcholine is exceeded, inhibition of secretion prevails. Why should excess of free acetylcholine lead to this effect? Welsh (1955) favours the suggestion of Zeller and Bissegger (1943) that a two-point attachment must occur between acetylcholine and its esterase: one between the positively charged nitrogen and a negatively charged group in the enzyme, and the other between the ester group and a group in the enzyme. Excess substrate may result in a single linkage to each molecule which prevents junctional transmission. On the other hand, the dual action of sympathomimetic amines on the blood vessels of a rabbit's ear was attributed by Burn and Robinson (1951) to changes in the availability of different receptors.

Further evidence may be found in the observations of Twarog (1954) on an invertebrate muscle which shows a reversal in response to repeated applications of a given concentration of adrenaline. The vasomotor reactions of the submaxillary gland (Stavraky, 1942; Graham and Stavraky, 1953 a, 1954 c), in which a prolonged infusion of adrenaline or injection of excessive quantities of this agent leads to vasoconstriction whereas small quantities cause vasodilatation, are in accord with this view.

In the case of the vascular response, the reversal may also be effected by a low level of the systemic blood pressure. The more ready occurrence of vasoconstriction in the submaxillary gland, when acetylcholine was injected during periods of low blood pressure, appears comparable to the enhancement of the vasoconstrictor effects of acetylcholine evoked by the reduction of the level of perfusion pressure reported by Alcock, Berry and Daly (1935) in the lung. A reduced rate of blood flow past the site of injection when the systemic blood pressure was low would presumably increase the concentration of acetylcholine in an organ. Recently Burton (1951) and Nichol, Girling, Jerrard, Claxton and Burton (1951) have shown that there is an increasing tendency towards instability on the part of small blood vessels as the pressure within them is reduced. It may well be that a lowered level of systemic blood pressure predisposes to the vasoconstrictor action of acetylcholine both through the factors which influence the concentration of this agent in the tissues of the gland and through the state of instability of the blood vessels concerned in the response.

The fact that partial deprivation of the internal organs of their innervation or excessive stimulation or both may lead to a reversal of the usual responses is of considerable practical significance and may have to be considered in many clinical derangements. The outstanding feature as far as the digestive glands are concerned is the finding that the body has the mechanism to regulate the composition of saliva and probably of other digestive juices with only one set of nerve fibres. This has actually been known for a long time. Babkin in 1913 divided the cervical sympathetic nerve in a dog with a chronic salivary fistula and obtained all the variations in the composition of saliva secreted in response to different foods as in normal animals. An explanation of this fact has so far been lacking. From the findings of Graham and Stavraky (1953 a, 1954 c) it appears that by changes in the quantity of liberated acetylcholine the glands may regulate vasomotor responses without the participation of the sympathetic nervous system and thus may alter the concentration of various constituents of the saliva, and possibly of other digestive juices, by the control of available fluid as well as by a specific secretory effect on the glandular cells.

2. Reversals and the Somatic Nervous System

Studying excitation, inhibition, and the effects of anaesthetic agents on excitability in somatic nerves, Wedensky (1903) described the phenomenon bearing his name. He noted that when appropriate repetitive stimulation of a peripheral motor nerve resulted in a tetanus of the muscle, additional stimuli caused a paradoxical decrease in the force of muscular contraction. Reduction of the intensity or frequency of the sustained stimulus had the opposite effect, causing an increase in the force of contraction of the muscle. Wedensky postulated that stimulation of the nerve leads to the development of a stable, non-propagated state in the nervous tissue which he called "parabiosis." He suggested that every excitatory process may be inhibited by a critical intensification of this state and thought that this phenomenon may have broad implications in the function of the nervous system. While later work indicated that "Wedensky's Inhibition" may be associated with a reduction of membrane potential in the nerve fibres and in the muscle, and is accompanied by changes in the threshold of their activation (see Thesleff, 1960), no completely satisfactory explanation of the phenomenon exists at the present time.

Would grading of the quantity of excitatory impulses regulate the activity of central neural mechanisms in a manner similar to the one seen in peripheral structures and would this regulation be influenced by partial denervation? Only a general survey involving the simplest types of experiments has, so far, been undertaken in order to answer these questions, but the results seem to warrant more work and will be presented with this in mind.

Fröhlich (1909) and Croisier (1944) found that weak ipsilateral stimulation of an afferent nerve caused contraction of the opening muscles of the claw of arthropodes, whereas strong stimulation led to a relaxation of the muscle, the opposite relation prevailing in the closing muscle of the claw. Sherrington and Sowton (1911 a, b, c), recording from the limb responses of the isolated quadriceps in the decerebrate cat, found that strong electric stimulation of an ipsilateral afferent nerve caused flexion, while weak stimulation of the same nerve resulted in extension of the limb.

In his studies on reciprocal innervation, Graham Brown (1911 a, b, 1912) and Brown and Sherrington (1912) also often encountered what Brown termed "irregular reactions." Recording the responses of the tibialis anticus and of the gastrocnemius soleus muscles in decerebrate cats, Brown observed that, as the intensity of ipsilateral afferent stimulation of a cutaneous nerve (saphenous) was increased, the response often changed from extensor contraction to co-contraction of the flexor and extensor and, finally, to reciprocal extensor relaxation accompanied by flexor contraction. Similar results were obtained on stimulation of the contralateral nerve. Brown could offer no explanation of this reversal and concluded that "neural balance" seemed to be disposed towards one response or another.

These almost forgotten observations were confirmed and expanded by Hinsey, Ranson and Doles (1930) and by Ranson and Hinsey (1931) who described a reversal of the crossed extensor reflex brought about by changes in the intensity and frequency of afferent stimulation and established instances of co-contraction of the muscles in spinal as well as in decerebrate animals. Ranson and his co-workers (Hinsey, Ranson and Doles, 1930, and Ranson and Hinsey, 1931) showed that these responses could be obtained on stimulation of posterior tibial and femoral nerves as well as on stimulation of the saphenous nerve. They stressed the relation between an increase in the intensity of afferent stimulation and the change from extensor to flexor contraction and concluded that both reciprocal responses and co-contractions of antagonistic muscles must be regarded as normal responses in intact limbs. Ipsilateral extension in response to afferent stimulation is characteristic not only of the hind limb, but Denny-Brown and Liddell (1927) and Miller (1934) described it in the fore limb, and other instances of such a reflex reversal became known. Denny-Brown and Liddell (1927), Bernhard and Skoglund (1942) and Fulton (1949, p. 218) suggested that the extensor reflex brought about by stimulation of ipsilateral nerves, the extensor thrust and the positive supporting reaction of Magnus (1924) were all phases of the same proprioceptive reflex elicited under different conditions. However, Hagbarth (1952, 1953) has recently confirmed the fact that cutaneous stimulation can yield ipsilateral extension, and showed that the nature of this response was determined by the skin area of the limb to which stimulation was applied, individual muscles being excited from specific cutaneous regions. Whether ipsilateral extension produced by such afferent stimulation is mono- or polysynaptic in nature is not definitely established. Lloyd (1943 a) found that afferent fibres of the size range having direct connection with motor neurones are mostly located in nerves of muscular origin, whereas fibres of the sizes which make connections with interneurones are predominantly found in cutaneous nerves. During this study and in the review of the functional organization of the spinal cord, Lloyd (1943 b, 1944) pointed out the lack of exact knowledge of the nature of the extensor thrust and of the crossed extensor reflex, and concluded that there was good reason for believing that they were both transmitted through multineurone pathways.

Cannon, Rosenblueth and García Ramos (1945) found that sensitization by chronic denervation of spinal neurones exaggerated reflex responses as well as reactions to chemical stimulating agents. It seemed interesting to see how chronic denervation would affect the reflex patterns of the irregular type described by Graham Brown (1911 a, b, 1912) and by Ranson and his co-workers (Hinsey *et al.*, 1930, and Ranson and Hinsey, 1931). In a study undertaken by Hughes, Stavraky and Teasdall (1950) and Hughes (1950), semidecerebration was carried out in cats four to six weeks before the experiment. Contractions of both quadriceps muscles were recorded, isotonically, in spinal or decerebrate preparations, in response to stimulation (square-wave) of the peroneopopliteal or posterior tibial nerve, repetitive stimulation being applied to the nerves through isolation transformers. As seen in the diagrammatic summary of the results shown in Fig. 35, ipsilateral contraction of the quadriceps in the control spinal animals was caused by weak stimuli, while higher voltages evoked co-contraction of the two muscles. In the decerebrate control cats, ipsilateral contractions, co-contraction of the two muscles, and relaxation of the ipsilateral quadriceps with a contraction of the opposite muscle were usually observed with a progressively increasing intensity of afferent stimulation. These results confirmed the findings reported by Ranson and his co-workers (Hinsey *et al.*, 1930, and Ranson and Hinsey, 1931). The critical current intensity varied widely in individual experiments, but the sequence described above was facilitated by the application of square-wave stimulation, which gave more reliable results than had been obtained with the induction coils used in the older investigations.

SPINAL CAT

RANGE OF STIMULUS	SPINAL CAT (NORMAL)			SPINAL CAT SENSITIZED BY PREVIOUS SEMIDECEREBRATION					
				AFFERENT STIMULATION ON SIDE OPPOSITE SEMIDECEREBRATION			AFFERENT STIMULATION ON SIDE OF SEMIDECEREBRATION		
	M.T.I. (VOLTS)	CROSSED RESPONSE	IPSILATERAL RESPONSE	M.T.I. (VOLTS)	CROSSED RESPONSE	IPSILATERAL RESPONSE	M.T.I. (VOLTS)	CROSSED RESPONSE	IPSILATERAL RESPONSE
FREQUENCY 10–60/SEC. PULSE DURATION .5 MSECS. 0.1 VOLTS →	0.3	—	∧	0.1	—	∧	0.3	—	∧
	0.6	∧	∧	0.7	∧	∧	0.9	∧	∧
		∧	∧	2.3	∧	∨	1.5	∨	∧
60 VOLTS		∧	∨	21.3	∧	∧	3.0	∨	∨

DECEREBRATE

RANGE OF STIMULUS	DECEREBRATE (NORMAL)			DECEREBRATE CAT SENSITIZED BY PREVIOUS SEMIDECEREBRATION					
				AFFERENT STIMULATION ON SIDE OPPOSITE SEMIDECEREBRATION			AFFERENT STIMULATION ON SIDE OF SEMIDECEREBRATION		
	M.T.I. (VOLTS)	CROSSED RESPONSE	IPSILATERAL RESPONSE	M.T.I. (VOLTS)	CROSSED RESPONSE	IPSILATERAL RESPONSE	M.T.I. (VOLTS)	CROSSED RESPONSE	IPSILATERAL RESPONSE
FREQUENCY 10–60/SEC. PULSE DURATION .5 MSECS. 0.1 VOLTS →	1.5	—	∧	0.3	—	∧	0.2	∧	∧
	2.0	∧	∧	0.8	∧	∧		∧	∧
		∧	∧	3.8	∧	∨		∧	∧
					∧	∨	9.1	∧	∨
60 VOLTS	17.5	∧	∨				11.2	∨	∨

— no response. ∧ contraction of the quadriceps. ∨ relaxation of the quadriceps. (Hughes, Stavraky and Teasdall, 1950)

Sensitization of the corresponding spinal centres by preceding semi-decerebration predisposed these centres to reflex reversal. Stimulation of the afferent nerve on the sensitized side resulted in ipsilateral inhibition in response to currents that were weaker than those applied to the control side. Thus, identical stimulation of the corresponding nerves on the two sides often led to reverse patterns of responses, the quadriceps ipsilateral to the semidecerebration contracting while the one on the sensitized side relaxed. Stronger currents caused a relaxation of both the quadriceps muscles in both the spinal and the decerebrate preparations (Figs. 35 and 36).

The fact that sensitization by means of a preceding semidecerebration greatly facilitated the occurrence of inhibition on the corresponding side of the spinal cord seemed to indicate that the effect was brought about through a central mechanism and not by the stimulation of different sets of nerve fibres in the peripheral nerve—a possibility considered by Sherrington and Sowton (1911 a, b, c). Other observations supported the view that the effect was of a central origin. For instance, a current intensity could be found which in a decerebrate cat produced inhibition of the quadriceps on the stimulated side. The cat was then converted into a high-spinal preparation, and when the same stimulation was repeated, the inhibition was replaced by an ipsilateral contraction of the quadriceps. This indicates the importance of bulbar facilitatory influences in the initiation of ipsilateral flexion as well as of extension. In an experiment carried out by Seguin (1956) impulses from the muscle itself produced the same effect. A strength of current was found which, when applied to the tibial nerve in a decerebrate preparation, caused bilateral contraction of the quadriceps muscles. When the same strength of current was applied on a background of proprioceptive discharge induced by weighting with 200 grams, the responses of the ipsilateral quadriceps were augmented. Stimulation when the muscles were weighted with 500 grams resulted in a relaxation of the ipsilateral quadriceps but contraction of the opposite one. The thresholds for all these effects were again

FIG. 35. Diagrammatic representation of reflex responses of both quadriceps muscles on electrical stimulation of the peroneo-popliteal or posterior tibial nerve with increasing current intensities in ordinary spinal and decerebrate cats and in similar preparations after semidecerebration four to six weeks before the experiment. (On the small sketch of the nervous system the horizontal line indicates the level of transection at the beginning of the experiment while the area in black shows the side of the preceding semidecerebration. The nerve and the electrodes are shown in relation to the semidecerebration and to the responses obtained.) M.I.T.: "mean threshold intensity" of stimulating current. Crossed response: response of the quadriceps contralateral to the stimulated nerve. Ipsilateral response of the quadriceps on the side of stimulation.

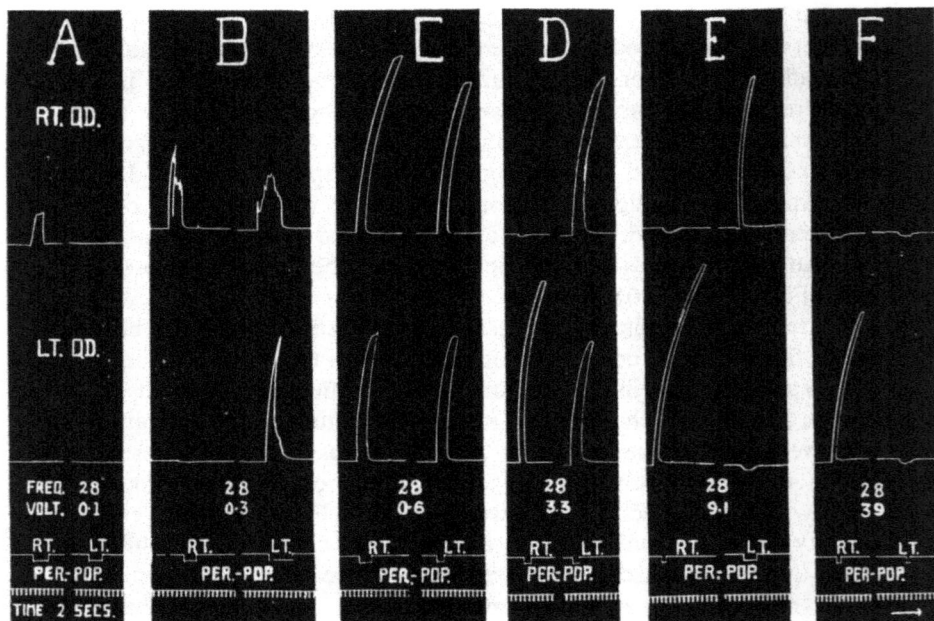

FIG. 36. Responses of the quadriceps muscles to increasing intensities of afferent stimulation (peroneo-popliteal nerve) in a decerebrate cat forty-one days after left semidecerebration. Note lower threshold of the right side (*A*) and a spread of excitation to that side with a stimulation of the left peroneo-popliteal nerve (*B*) in the absence of a similar response contralaterally; the reversal from contraction to relaxation is also obtained first on that side (*D*), as well as the spread of inhibition with contralateral stimulation (*F*).
(Hughes, 1950)

lower on the partially isolated side of the spinal cord. These experiments show that reflex reversal from excitation to inhibition can be produced by a summation of excitatory influences when these latter reach a critical density. As the change from excitation to inhibition occurred with the application of a given electrical stimulation to the peripheral nerve, it invoked the same afferent fibre spectrum in each case; thus, the possibility of a predominant stimulation of excitatory or inhibitory fibres in different experimental situations was excluded.

In another investigation (Graham and Stavraky, 1953 b, 1954 a, b; Graham, 1954), the effects of sensitization by partial isolation were studied in relation to reciprocal innervation. Muscle recordings from antagonists in the thigh and shank of the cat were made bilaterally and effects of chronic semidecerebration or semisection of the spinal cord on reflex patterns were studied in decerebrate and in low-spinal cats. The pattern of responses in the conventional decerebrate preparation with increasing stimulus intensity changed from extensor contraction to co-contraction of the extensor and flexor, and to extensor relaxation accompanied by flexor contraction (reciprocal innervation). The complete sequence was observed in the reaction of the shank antagonists, first ipsilateral to nerve stimulation, and then, with very intense afferent stimulation, also on the opposite side. The thigh antagonists seemed less sensitive than the shank muscles to the stimulation of the posterior tibial or superficial peroneal nerve used in this study. Here, too, the complete sequence of responses was observed on the side of the stimulation, but only the first two phases of the pattern—the extensor contraction and co-contraction occurred contralaterally.

When the experiments were repeated in semidecerebrate animals (13 to 228 days after the operation), essentially the same pattern of responses was observed with increasing intensity of afferent stimulation, but the thresholds for all these effects were lower on the side opposite the previous semidecerebration. In fact the amplitude of the contractions was often greater on that side. If this was not the case, as during the extensor contraction contralateral to nerve stimulation, cutting of the nerves to the flexors brought out the response. Static loading of the individual muscles with light weights generally increased the amplitude of the contraction; heavy weights reduced it. The contraction of antagonists was reduced or abolished concurrently, indicating a spread of the inhibition. In these semidecerebrate cats, the inhibitory effects of weighting on extensor reactions were particularly prominent contralaterally to the semidecerebration.

The central state of excitation was also altered in other ways. Spiegel and Scala (1943) found that galvanic currents applied to the vestibular apparatus caused alterations in decerebrate rigidity. They observed this phenomenon in the intact fore limbs of decerebrate cats, anodic stimulation causing apparent reduction in decerebrate rigidity, while cathodic stimulation caused augmentation of the extensor tone of the fore limbs. When applied by Graham (1954) to animals sensitized by a preceding semidecerebration, anodic stimulation caused a complicated sequence of changes in tone of the hind limbs. Cathodic stimulation usually produced in the muscles of the shank an extensor contraction when currents of a low amperage were applied; with greater amperages, a co-contraction of the antagonistic muscles took place, the effects being more pronounced on the chronically decentralized side. In another experiment, a subliminal afferent stimulation was added in an ordinary decerebrate cat to the effect of weak galvanic stimulation of the vestibular apparatus. The latter

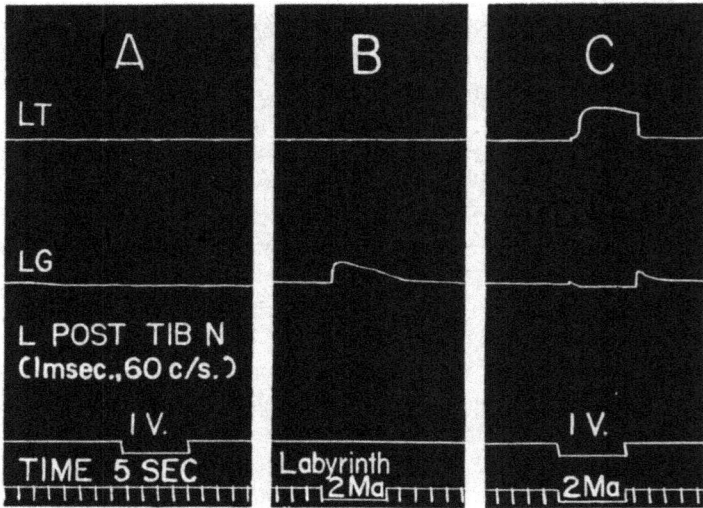

Fig. 37. Summation of effects of limb-afferent and vestibular stimulation in decerebrate cat. *A*: subliminal square wave stimulation (one volt) of left posterior tibial nerve. *B*: galvanic stimulation (two milliamperes) of the labyrinth (right ear electrode negative: left, positive). *C*: stimulations A and B combined. Note absence of either contraction or post-inhibitory rebound in *A*; contraction of the left gastrocnemius muscle (LG) but no effect on the left tibialis muscle (LT) in *B*; inhibition of the gastrocnemius with a post-inhibitory rebound and contraction of the tibialis in *C* when the two stimulations were combined. (Graham, 1954)

stimulation, before the addition of the stimulation of the ipsilateral nerve, had caused a weak contraction of the gastrocnemius muscle. When the stimulations were combined, the two trains of nerve impulses produced an inhibition of the extensor tone and the appearance of a flexor contraction which neither had accomplished separately (Fig. 37). The recordings in these experiments were made with the application of minimal tension to the muscles, in order to avoid, as much as possible, activation of the muscle proprioceptors. This precluded the gastrocnemius from relaxing below the resting level during the combined stimulation, but the postinhibitory rebound in the muscle indicated the extent of inhibition which took place. Similar results were obtained with weighting of the muscle: application of loads to the gastrocnemius caused the reaction to galvanic stimulation of the labyrinth to change from contraction to relaxation. It is interesting that Gernandt and Terzuolo (1955 a, b) made similar observations with a different approach. Recording strychnine-induced potentials in the spinal cord of curarized, decerebrate cats, they found that the electrical activity could be inhibited by appropriate changes in the position of the head of the animal and that this effect was reversed by section of the spinal cord below the recording electrodes and by cerebellectomy.

In low-spinal preparations semisection of the spinal cord was performed by Graham and Stavraky (1953 b, 1954 a, b) and Graham (1954) at T-12 level 10 to 660 days before the experiment. At the time of the experiment, the animals were decerebrated and the section of the cord completed one segment below or above the chronic lession. In control cats transected acutely, flexor contraction or co-contraction of the flexor and extensor prevailed both on the side of the afferent stimulation and contralaterally. In previously semisected animals, crossed and ipsilateral reflexes were elicited more readily on the side of chronic semisection of the cord, the amplitude of the contraction in the case of submaximal stimulation being greater on that side. Also, with increasing intensity of stimulation of the posterior tibial or superficial peroneal nerve, reciprocal flexor contraction and extensor relaxation were observed on the side of chronic semisection. These results indicated again that chronic semisection of the spinal cord brings out group reactions of antagonistic muscles and facilitates the re-establishment of reciprocal innervation when it occurs on the decentralized side. Thus, the reciprocal effect had the appearance of a reversal in previously isolated areas of the spinal cord, while on the control side a corresponding stimulation produced usually only excitatory effects.

The influence of pentylenetetrazol and strychnine on reflex reversals of partially isolated regions of the spinal cord were studied by Seguin (1956) in frontal-lobectomized and semidecerebrated cats. At the time of the experiment the animals were decerebrated, and the effects of stimulation of the posterior tibial nerves on weighted quadriceps muscles (200–500 gm.) were recorded. With increasing intensities of afferent stimulation the ipsilateral response of the quadriceps changed from contraction to relaxation, the thresholds for both these responses being lower on the previously isolated side. Pentylenetetrazol (5–25 mgm./kgm.) and strychnine (0.01–0.1 mgm./kgm.), injected intra-aortally, increased the range of stimulation intensities which brought about contractions. This effect was produced with smaller doses of the drugs and was more marked on the side opposite the chronic cerebral ablation. The threshold for both contraction and relaxation of the quadriceps remained lower on that side than on the control one. The crossed extensor reflex was influenced in a more complicated way than other responses by these drugs, the effects depending on the dosage of pentylenetetrazol and strychnine used and on the strength of the stimulus,

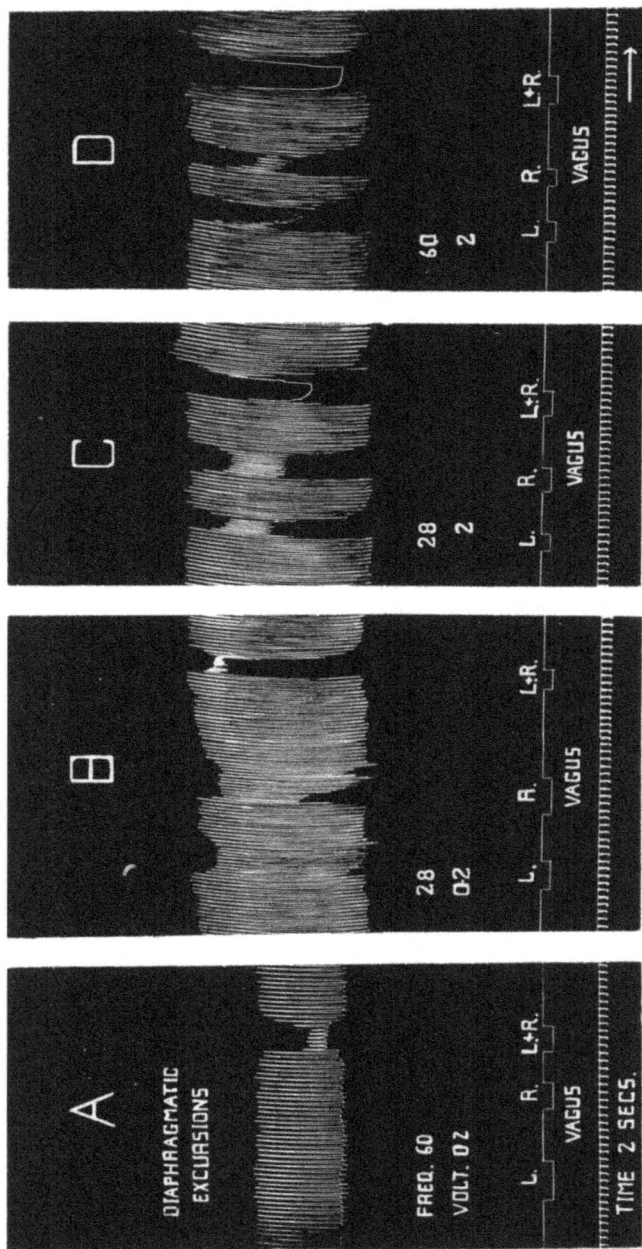

FIG. 38. Central summation of the effects of stimulation of the vagi nerves in the neck in the rabbit recorded by means of a "diaphragmatic slip." *A*: stimulation of the vagi nerves separately is ineffective, but effective summation takes place when both nerves are stimulated simultaneously (L. & R.). *B*: summation of effects of minimal stimulation results in an acceleration of the respiratory movements and in greatly reduced excursions, almost an arrest "in inspiration." *C* and *D*: application of progressively stronger currents results in an acceleration of respiratory movements when the vagi are stimulated separately on each side, but in an expiratory arrest when they are stimulated together, the sixty cycles per second frequency in *D* being more effective than the twenty-eight cycles per second in *C*. (Hughes, Stavraky and Teasdall, 1950)

but all the observed effects were greatly enhanced by preceding isolation of the spinal neurones. It is interesting that, when large quantities of strychnine (0.1 mgm./kgm.) or pentylenetetrazol (25 mgm./kgm.) resulted in bilateral clonic contractions of the quadriceps muscles, these twitches could be completely inhibited by afferent stimulation of either posterior tibial nerve. This again shows that convergence of excessive excitatory influences on spinal motor neurones results in inhibition.

Inhibition of reflex action was also brought about by the injection of pentylenetetrazol into the brain stem through the vertebral arteries. The effect was more pronounced with the crossed than with the ipsilateral response. The difference between the action of strychnine and of pentylenetetrazol on crossed responses was attributed to the prominence of action of pentylenetetrazol on midbrain structures.

Wyss (1940, 1944, 1947 a, b) was able to bring about reflex reversal of respiratory movements by the application of stimuli of increasing frequency or by a change from an in-phase to an out-of-phase stimulation of the vagus nerves. Hughes, Stavraky and Teasdall (1950) reproduced the central summation of excitation seen by Wyss by stimulating alternately one vagus nerve and then both vagi with currents of the same frequency, duration and intensity. Currents, which were subliminal when applied to each vagus nerve separately, became effective and caused acceleration of breathing when applied to both vagi simultaneously (Fig. 38). When stronger currents, ranging in intensity from one to six or seven volts, were applied to the central end of each vagus nerve separately, they produced an inspiratory arrest and, sometimes, a response which consisted of a reduction in amplitude and an acceleration of the respiratory movements. Applied to both vagi simultaneously, the same currents caused an arrest of breathing in expiration.

The development of various types of inhibition (extinction, differentiation, delay, etc.) described by Pavlov (1928) and his associates during the study of "conditioned reflexes" may be regarded as evidence that processes analogous to those occurring in the spinal cord and in bulbar structures take place in the highest levels of the central nervous system. These studies are widely known but, because they form a special field of physiology, are beyond the scope of the present discussion.

3. Mechanism of Reversals and Their Relation to Supersensitivity

It is tempting but undoubtedly premature to postulate that excessive quantities of neurochemical transmitters cause reversals at central synapses similar to the effects which acetylcholine or adrenaline have in peripheral structures under the control of the autonomic nervous system. On the other hand, to adopt a general theory based on the phenomenon of "Wedensky's Inhibition" (1903) in the present state of our knowledge would necessitate embracing a descriptive view which involves an abstract concept. Possibly for this reason attempts at explanations based upon a morphological visualization of distinct excitatory and inhibitory sets of fibres and receptors have often been favoured.

The importance of muscle and tendon proprioceptors in the regulation of reflex muscular activity has been emphasized since the time of Sherrington (1894) and there can be no doubt that these peripheral sensory organs participate in many reflex reversals which were described in this chapter. The analogy of the results of weighting the muscles in the experiments of Graham (1954) and Seguin (1956) with the findings obtained during the electrophysiological analysis of the effect of stretch receptors on reflex activity by Matthews (1933), Hunt and Kuffler (1951 b), Hunt (1952 b) and others is quite striking.

The influence exerted by proprioceptive organs on motor activity can be either excitatory or inhibitory. This led early in the study of the role of muscle spindles and tendon organs to a divergence of opinion as to their part in the regulation of muscular contractions. This divergence culminated in the theory of Fulton and Pi-Suñer (1928) on the one hand, and that of Denny-Brown (1928) on the other (see Lissman, 1950). Subsequent electrophysiological studies pertaining to the effect of muscle and tendon proprioceptors expanded the concept of a peripheral regulation of motor activity. The discovery by Leksell (1945) of the control of tension in the muscle spindle itself by the small motor fibres known as the gamma-efferent system gave impetus to a great volume of work which was carried out with intricate and varied techniques, but which often resulted again in different concepts regarding the role played by the muscle and tendon receptors (see pages 66–67). The fact should be emphasized that both facilitation and inhibition may be obtained on the activation of proprioceptors, the trend in general seeming to be in the direction of an initial facilitation which with more marked tension in the muscle or with stronger stimuli is changed to inhibition. As far as the muscle spindle is concerned, this effect is usually accounted for by the participation of different receptors—annulo-spiral and flower-spray. The low-threshold receptors supposedly cause facilitation and the high-threshold ones bring about inhibition (see Granit, 1955, Fig. 109, p. 235). However, in the Golgi tendon organs, only one type of receptor is known to be present; yet, as shown by McCouch *et al.* (1950), electric stimulation of the quadriceps tendon may yield an early facilitation which is followed by inhibition. These and other experiments led Granit (1955) to suggest the possibility of the existence of another as yet undetermined recèptor. In his concluding remarks Granit (1955, p. 275) states: "The tendon jerk is . . . quite capable of surprising the experimenter just as much as its nearest electrical equivalent, the monosynaptic reflex. In both cases it is best to admit that we do not yet understand all aspects of what constitutes motoneurone excitability."

If there is little doubt that some proprioceptors are predominantly excitatory and others inhibitory under certain conditions of observation, the nature of the influence exerted by their central connections is obscure. Furthermore, the experiments summarized in this chapter show that cutaneous afferents also regulate the reflex response. But what is even more significant, these experiments bring out the fact that the nature of the response is regulated by the central excitatory state. The functional grouping of the afferent impressions and the state of the neuronal pools, rather than the anatomical entities, determine the response. Many facts bear out this view: (1) Simultaneous activation of several converging pathways commonly leads to an inhibition of a neurone pool whereas, when stimulated separately, these paths cause excitation. (2) Facilitation of a group of neurones by preceding excitation causes a stimulus, which ordinarily exerts an excitatory effect, to produce inhibition. (3) An increase in the intensity of stimulation applied to an afferent nerve usually causes a change from excitation to inhibition; in some instances a change in the opposite direction may be produced. These effects can be seen on stimulation, not only of branches which contain fibres from muscle proprioceptors, but also when purely cutaneous nerves are stimulated. (4) Increased neuronal excitability, which results from the introduction of chemical agents such as strychnine and pentylenetetrazol, changes the effect of stimulation of an afferent nerve. Because the site of action of these drugs is in the central nervous system and because the electrical stimulation involves the same nerve-fibre spectrum as before, the reversal must be attributed to some change in the neurones upon which the nerve impulses impinge. (5). Neuronal supersensitivity induced by partial denervation greatly increases the tendency to reflex reversal, thus emphasizing once more the central nature of the process which determines whether the neuronal pool will respond to a given train of impulses by excitation or by inhibition. At the same time, the increased excitability which is brought about by partial isolation of different areas of the nervous system results in a greater ability to co-ordinate individual reflexes into group reactions which involve reciprocal innervation. This is true both in spinal and in decerebrate animals.

Several views as to the nature of central reversal may be mentioned. All of them are basically associated with the problem of inhibition. The suggestion of Wyss (1940, 1944, 1947 a, b) that the reflex reversal, caused by changes in frequency of the stimulating current, depends upon the presence of intercalated neurones of various thresholds in the central nervous system, if true, should be applicable in different intensities of afferent stimulation and in other conditions. Acceptance of this explanation, however, would necessitate assuming the existence of several sets of intercalated neurones, one set responding to certain frequencies, another to certain intensities of stimulation, and so on. Besides this, such an explanation would not readily account for reflex reversals caused by sensitization by means of partial denervation. Moreover, the reflex reversals caused by various chemical agents and by fatigue would be difficult to interpret on the basis of this hypothesis.

Another explanation of reflex reversal as it applies to "reciprocal innervation" was suggested by Gasser (1937), who visualized the arrival of excitatory and inhibitory impulses through different fibres to several intercalated neurones. According to Gasser's view, the numerical preponderance of discharging synaptic terminations on these intercalated neurones and the consequent synchrony or asynchrony of firing of the latter determine whether the ultimate effect will be excitation or inhibition of the motor neurones.

The concept of inhibition as a block in interneuronal pathways was supported by Bremer and Bonnet (1942) and undoubtedly plays a prominent part in multisynaptic reflexes, but the demonstration of direct inhibition in two-neurone paths (Lloyd, 1946 a) resulted in the revival of interest in mechanisms which would not necessitate the involvement of a state of subnormality in interneurones. Recent trends of thought in this respect cover several possibilities: (1) Barron and Matthews (1935, 1938) and Renshaw (1946) favoured the view that inhibitory impulses block presynaptic terminals. (2) Brooks and Eccles (1947) invoked the Golgi cell theory which supposed that inhibitory impulses created anelectrotonic foci on the surface of motor neurones. (3) Gesell (1940), Gerard (1941) and Dontas and Gesell (1955) postulated the existence of a fluctuating potential difference between the dendrites and axon hillock of a neurone. Impulses arriving at the latter may result in inhibition by diminishing the resting potential difference. However, the structural basis for this mechanism of inhibition was actually demonstrated only in the Mauthner's cells of lower vertebrates (see Retzlaff, 1954, 1955).

Lately the possibility of a chemical transmission of inhibitory impulses acquired considerable support from Eccles (1953), Florey (1953,

1954 a, b, 1956), Eccles, Fatt and Koketsu (1954), Pfeifer and Pataky (1955), Pataky and Pfeifer (1955), Florey and McLennan (1955 a, b, c), Elliott and Florey (1956) and McLennan (1957). It is interesting to note that, as pointed out by McLennan (1957), extracts containing Factor I also may show excitatory effects which appear to be due to the

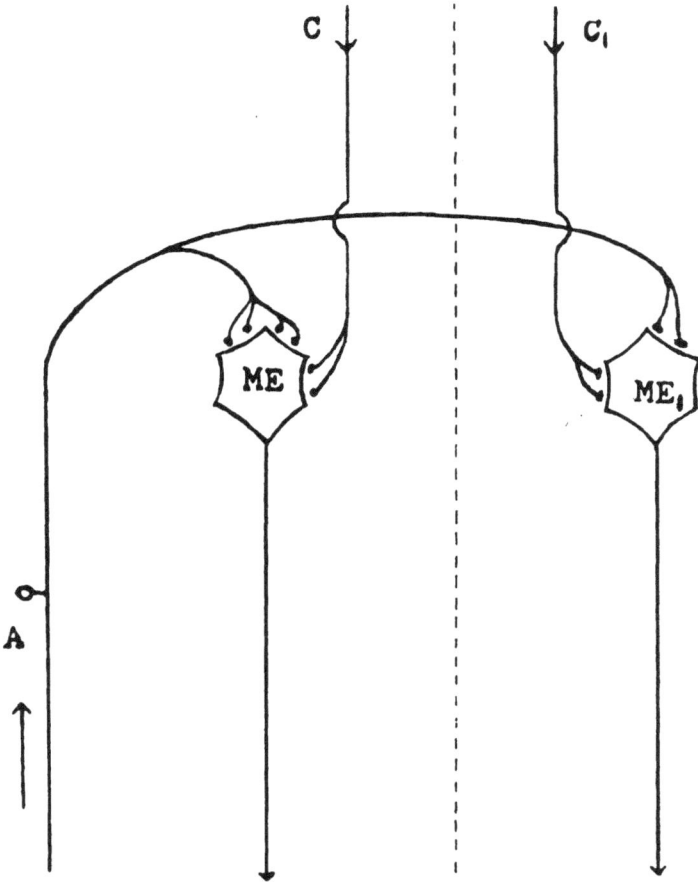

FIG. 39. A morphologic visualization of the proposed explanation of reflex reversals based on the variation in the number of impulses reaching the effector neurones (scheme omits interneurones and does not attempt to express actual numerical relation of end-bulbs to motor neurones). A: afferent fibre. ME: extensor motor neurone on the side of the stimulated afferent fibre. This neurone receives a larger number of terminal ramifications from the afferent fibre than the extensor motor neurone ME_1 situated on the opposite side of the spinal cord. C and C_1: descending fibres from higher levels of the central nervous system which may also send variable numbers of terminals to each side. (Hughes, Stavraky and Teasdall, 1950)

factor itself: the activities of the hypoglossal nucleus and flexor reflexes in cats may be enhanced (Florey and McLennan, 1955 c) and contraction of the squid rectum elicited (Florey, 1956).

A possibility which does not predetermine the intimate nature of synaptic transmission and which would cover instances of direct as well as indirect inhibition was put forth by Hughes, Stavraky and Teasdall (1950). They suggested that the number of nerve impulses which reach a given group of neurones in a given time determines the type of response of these neurones, provided other conditions remain unaltered. Accordingly, a certain optimal number of impulses would lead to excitation while an excessive number of impulses would result in inhibition (see Fig. 39). Basically this view is related to the one expressed by Wedensky (1903) and to that adopted by the Pavlovian School of Physiology (Konradi, 1958, pp. 543–554).

While applicable to many instances of inhibition this scheme may also account for some of the more complex changes seen in multi-neuronal reflexes. For instance, the general trend with progressively increasing intensity of stimulation of an afferent nerve or with summation of different excitatory effects is usually the change from excitation to inhibition. However, if after this effect is attained, the intensity of stimulation—that is, the number of nerve fibres involved and the number of impulses converging on the neuronal pool—is further increased, the inhibitory effect occasionally disappears and excitation breaks through once again, as judged by the resulting reflex response (Hughes, Stavraky and Teasdall, 1950; Hunt, 1952 b; and others).

A similar situation prevails when the effects of afferent stimulation before and after the injection of strychnine or pentylenetetrazol are compared (Sherrington, 1905 a, 1906, 1907; Owen and Sherrington, 1911; Knoefel and Murrell, 1935; Bremer, 1944; Bradley and Schlapp, 1950; Bradley, Easton and Eccles, 1953; Brooks, Curtis and Eccles, 1955; Seguin, 1956), as well as in other experimental conditions such as weighting of the muscles (Hunt, 1952 b; Graham, 1954; Seguin, 1956). It is conceivable that in these instances of the return of the positive reflex response an overexcitation of internuncials results in their inhibition, thereby diminishing the barrage of impulses arriving at the motor neurones. This would relieve the latter from overexcitation and allow them to resume their activity.

On the basis of this hypothesis it is conceivable that there are peripheral receptors and groups of neurones which send densely distributed terminations to certain regions of the central nervous system and which lead predominantly, or even exclusively, to inhibition; and others which send dispersed systems of terminations and which cause excitation unless assisted by a train of impulses arriving by way of another system of nervous connections.

A number of reflex reversals mentioned in the literature, which are caused by the position of the head and neck (Socin and Van Leeuwen, 1914; Beritoff, 1915; Girndt, 1926; Pollock and Davis, 1924; Pi-Suñer and Fulton, 1929; and others) and by the posture of the limbs (Gergens, 1876; Sherrington, 1900, 1905 a, b; Von Uexkull, 1904; Magnus, 1909, 1910; Brown, 1911), may be accounted for in this manner.

The greater ease of reflex reversal which takes place after sensitization by partial denervation (Hughes, Stavraky and Teasdall, 1950; Teasdall and Stavraky, 1953, 1955; Seguin, 1956; Graham, 1954) falls also into the group of responses in which a greater effectiveness of the excitation leads to inhibition. The latter is achieved, however, not by an increased or reduced number of impulses converging upon the neurones, but by a change in the sensitivity of the neurones themselves to the arriving impulses. Deterioration in the general condition of the animal may lead to the opposite effect—the disappearance of inhibition and the predominance of an excitatory response, this again signifying a change in the state of the neurones which manifests itself in lowered irritability.

This view makes it possible to reconcile the co-operative action of the flexors and extensors, which is often seen in intact limbs, with the principle of reciprocal innervation demonstrated by Sherrington (Creed *et al.*, 1938, pp. 38–40) on isolated groups of muscles, the two responses possibly depending upon different degrees of excitation. Similarly, some instances of "rebound" may be thought of as a contraction caused by a subsiding excitatory process which, during the actual stimulation, was strong enough to cause a reversal in all or some of the neurones, these neurones during the resolution of the hyperexcitation going through a stage of active discharge.

On the clinical side, the state of augmented sensitivity may play a part in the conversion of the physiological extension (downward movement of the large toe), which occurs when the sole of the foot is stroked in normal human beings, into a physiological flexion—also called "dorsiflexion" of the large toe in patients with upper motor neurone lesions (Fulton and Keller, 1932, pp. 3–14). Another possible example of a reversal, this one of a strictly physiological type, may be given: the general "wilting" that takes place in extreme fright and is very marked in certain species of animals, such as the rabbit and in some birds, and is also known to occur in man may be interpreted as being the ultimate result of overexcitation and a massive reversal of all the reflex activity.

Finally, it may be recalled that Pavlov (1941), in his studies on conditioned reflexes, by forcing the process of conditioning of dogs beyond the limits of physiological endurance noted a change in the responses of the animals which he ascribed to the development of a so-called "protective inhibition" ("equalization," "paradoxical" and "transmarginal" phases). Pavlov was inclined to regard these cases of augmented inhibition as manifestations of a single process intimately connected with excitation. The abnormal conditioned responses, once developed, were seen to persist for long periods of time and Pavlov expressed the belief that many neuroses and psychoses in human patients may have a similar basis. This view was recently considered by Sargant (1957) in relation to combat fatigue and some other states in which heightened suggestibility plays a part. Thus, apparently excessive and unduly prolonged stimuli can lead to reversals and to development of inhibition of cerebral cortical responses of the acquired type as well as of inborn reflexes.

The analogy of this general trend to the one seen in peripheral structures which are under the control of the autonomic nervous system is quite remarkable. A common principle seems to govern bodily functions in this respect; whether the intimate nature of the effect is the same in the central nervous system and in the peripheral organs is a question that cannot be answered at this time. However, it is interesting that stimulation of the chorda tympani was shown by Burgen (1956) to cause a loss of potassium from the salivary glands both into the saliva and into the plasma of glandular venous blood. This was accompanied by an increase in the concentration of cellular sodium, and a decrease in plasma sodium. When stimulation was stopped, the plasma potassium level fell and its concentration in the salivary gland slowly rose until normal conditions were restored. It may be no coincidence that a much more transient, but essentially similar, ionic shift occurs during the transmission of an impulse along a nerve fibre and in the neurone itself during its activation.

1. RESTITUTION OF FUNCTION FOLLOWING LESIONS OF THE CENTRAL
NERVOUS SYSTEM AND ITS RELATION TO DISINHIBITION, AUGMENTED
INFLUX AND SUPERSENSITIVITY DUE TO NEURONAL ISOLATION

All the presented evidence bears out the fact that excitability of living matter is subject to change. Severance of part of the nervous connections results in a depression of the excitability of an isolated region and is later followed by an increase. It was suggested by Cannon and Rosenblueth (1949, p. 206) that in the central nervous system this sequence of events may be the basis of spinal shock and of the subsequent partial recovery of some of the functions of the spinal cord after its section. In fact, Cannon and Rosenblueth (1949, p. 214) considered the possibility that supersensitivity brought about by partial isolation "is an important factor in so-called reorganization of the central nervous system after injuries" in general. A considerable degree of support can be mustered for such a view.

The occurrence of a progressive compensation for an anatomical defect was recognized by Hughlings Jackson (1884). As summed up by Pike, Elsberg, McCulloch and Rizzolo (1929, p. 260), Jackson attributed this property to a "change in the quantity of nervous energy flowing through a given mechanism or pathway or level." However, Jackson did not seem to have given any tangible idea as to how this increased quantity of nervous energy could be generated, and the vagueness of such a hypothesis made it one of the less celebrated features of his teachings.

On the other hand, as suggested by Sherrington (1898 a, 1906), Creed et al. (1938) and Monakow (1914), and later shown by Trendelenburg (1910), Fulton, Liddell and Rioch (1930), and by Liddell (1934, 1936), and endorsed by the evidence brought forth by Magoun and Rhines (1947, p. 34), spinal shock may be regarded as the result of the withdrawal of facilitatory influences descending into the spinal cord.

After transection of the cord, recovery of the reflexes gradually takes place. Liddell (1934), studying the susceptibility of the knee-jerk of cats to single-shock inhibition, observed that a restitution of the central excitatory state to almost normal values occurs within sixty days after section of the cord. He concluded that "even, if real, such findings are useless to the organism because the command from higher centers is permanently lost in a complete transection." In a similar preparation, Graham (1954) studied patterns of reflex responses which occurred with increasing intensity of afferent stimulation in antagonists around the ankle joint. Within nine days after transection of the cord he found a marked restoration of the more complex responses involving reciprocal innervation.

Even more striking recovery of function takes place after chronic semisection of the spinal cord. Observations on cats led Pike, Elsberg, McCulloch and Rizzolo (1929) to conclude that "some new mechanism which may have borne a part of the functional burden of control of movements of the limb when the cord was intact takes over the entire control of such movements." In agreement with this view, Asratian (1953, pp. 281, 282) expressed the belief that following semisection and other partial lesions in the spinal cord, a large degree of functional recovery was associated with the presence of the cerebral cortex and was abolished after its subsequent destruction.

A study of extensor and flexor reflexes in isolated thigh muscles of the cat was made by Cannon, Rosenblueth and García Ramos (1945) and Rosenblueth, García Ramos and Cannon (1945) in experiments carried out eight to twenty-five days after semisection of the spinal cord. They found greater reflex excitability on the side of the semisection as compared to the acutely transected side. This was true of the tendon jerks and of the after-discharge of the reflex response to ipsilateral and contralateral stimulation of the peroneo-popliteal nerve. Fulton and McCouch (1937) and McCouch, Hughes and Stewart (1943) described enhanced recovery of reflex excitability on the side of the chronic semisection in monkeys and baboons after complete transection of the cord.

Graham and Stavraky (1954 a, b) and Graham (1954) confirmed these observations and found, both in spinal and decerebrate preparations, a greater tendency towards organization into patterns involving reciprocal innervation on the side of the previous semisection. Teasdall and Magladery (1956) reported an enhancement of both early (monosynaptic) and late (polysynaptic) reflex potentials six to seventy-four days after semisection of the spinal cord at D5 to D10 level, and Liddell

(1936) demonstrated in decerebrate cats 200 to 300 days after a lesion in the ventrolateral area of the cord "overcompensation" both as regards the susceptibility of the knee-jerk to ipsilateral inhibition and "reflex resting tension" in the quadriceps.

Not only lesions of the spinal cord but also cerebral ablations, such as removal of the motor cortex, frontal lobectomy and semidecerebration, lead to an exaggeration of the excitability on the corresponding side in spinal cat preparations. Return of tendon reflexes in the whole limbs was observed in monkeys and chimpanzees by McCouch (1924), Fulton and Keller (1932) and by Fulton and McCouch (1937). Hughes, Stavraky and Teasdall (1950), studying the effect of increasing intensity of afferent stimulation in the isolated quadriceps muscles of the cat, observed an increased reflex excitability and an increased tendency to inhibition on the side opposite a semidecerebration of four to six weeks' standing as compared with the responses of the quadriceps on the previously intact side, and Graham and Stavraky (1954 a, b) and Graham (1954) noted a greater complexity of reflex patterns. The most conclusive evidence that the excitability of partially isolated spinal neurones is augmented after upper motor neurone lesions was presented by Drake, Seguin and Stavraky (1956). In cats and white rats, made spinal at various times after removal of one frontal lobe or after semidecerebration, recordings of the electrical activity of anterior horn cells and of the contractions of isolated muscles were made and the sensitivity of the two sides of the cord to pentylenetetrazol, camphor, picrotoxin and acetylcholine was measured. Up to sixteen days after the initial operation the responses were practically symmetrical on the two sides (Fig. 40). These results were comparable with the ones described by Sprague, Schreiner, Lindsley and Magoun (1948) and Schreiner, Lindsley and Magoun (1949), who saw that transection of the spinal cord abolished the hyperactivity of the stretch reflexes on the side of the previous cerebral ablation. However, three to fifteen months after the initial operation, the electrical discharges from the motor neurones were greater and the threshold was lower on the previously isolated side of the spinal preparations. The quadriceps on the side corresponding to the chronic cerebral ablation also responded to smaller doses of the injected drugs, and contracted more vigorously (Figs. 41 and 42).

Hyperexcitability of the side corresponding to the chronic cerebral lesion may be demonstrated in the decerebrate animal as well as in the spinal preparation. This was seen in the case of segmental afferent stimulation by Hughes, Stavraky and Teasdall (1950), Graham and Stavraky (1953 b, 1954 a, b), Graham (1954) and Seguin (1956) in

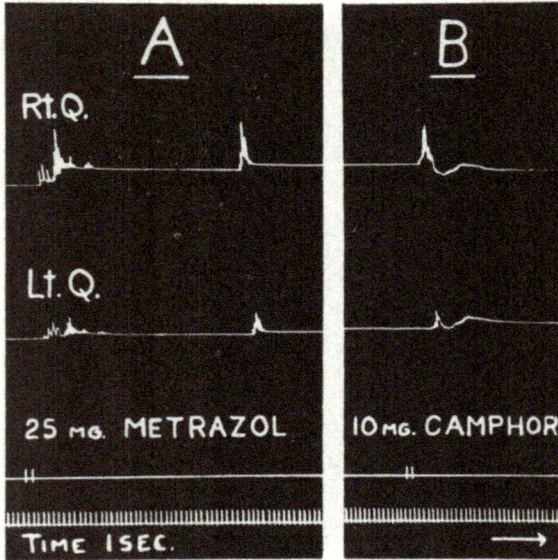

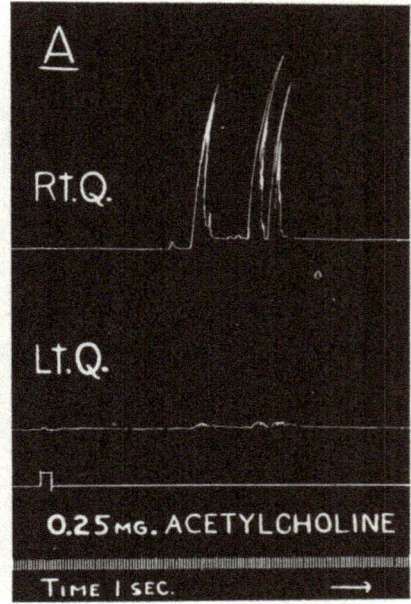

Fig. 40   Fig. 41(A)

Fig. 40. Isotonic myograms showing the effects of intra-aortal injections of pentylenetetrazol (*A*) and camphor (*B*) in a high-spinal cat sixteen days after left semidecerebration. Rt. Q.: right quadriceps, Lt. Q.: left quadriceps. Note practically identical responses on the two sides. (Drake, Seguin and Stavraky, 1956)

Fig. 41. Effects of intra-aortal injections of acetylcholine (*A*) and pentylenetetrazol (*B*) in a high-spinal cat 120 days after left semidecerebration (recording and abbreviations as in Fig. 40). Note marked exaggeration of the responses on the right side. (Drake, Seguin and Stavraky, 1956)

Fig. 42. Oscilloscopic tracing of the effects of an intra-aortal injection of acetylcholine recorded from the central end of severed femoral nerves in a curarized high-spinal cat fifteen months after a left frontal lobectomy (motor cortex included in the removal). Note practically no effect of 0.3 mgm./kgm. of acetylcholine on the control side, but a marked increase of the electrical activity on the side opposite the frontal lobectomy. (Drake, Seguin and Stavraky, 1956)

Fig. 43. Action potentials recorded from L7 ventral roots in a high-spinal cat (acute experiment) following semisection of the spinal cord at T8 level one month previously. Single shocks of decreasing intensity were applied to the proximal portion of severed tibial nerve in popliteal fossa. Left-hand column: intact side. Right-hand column: semisected side. *A*: monosynaptic responses. *B*: polysynaptic responses (see text). (Teasdall, Magladery and Ramey, 1958)

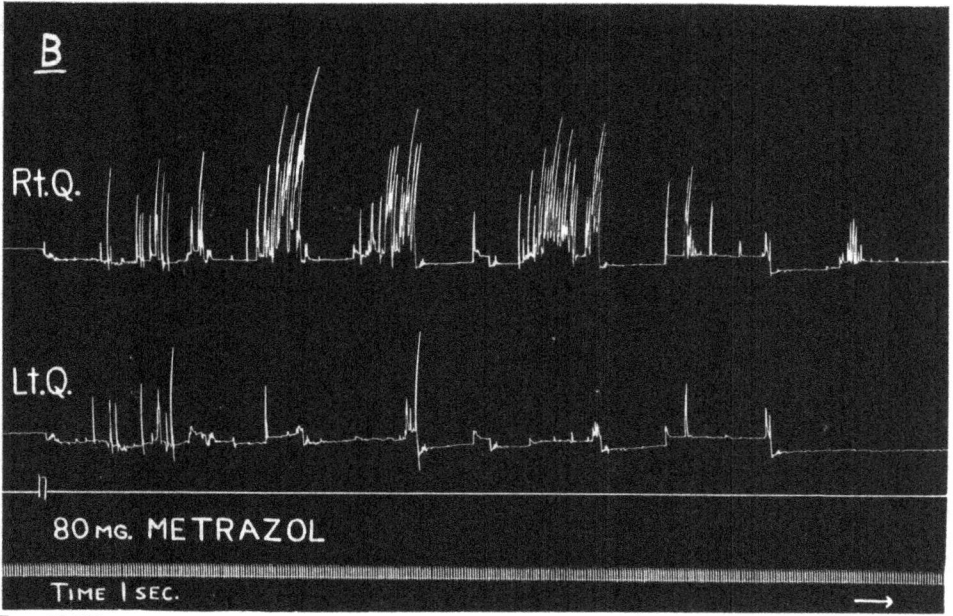

B

Rt.Q.

Lt.Q.

80 MG. METRAZOL

TIME I SEC.

FIG. 41(B)

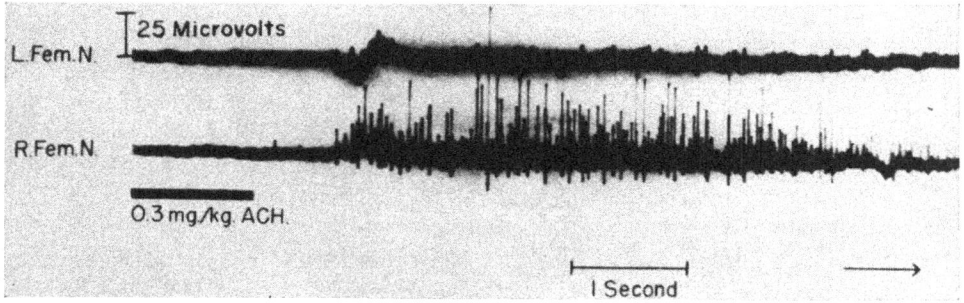

25 Microvolts

L.Fem.N.

R.Fem.N.

0.3 mg/kg. ACH.

I Second

FIG. 42

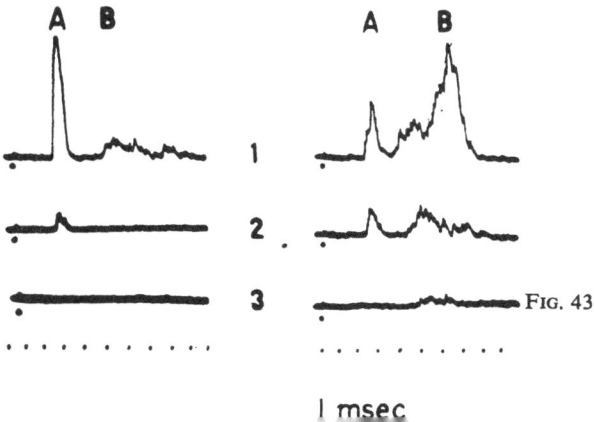

A B   A B

1

2

3   FIG. 43

I msec

experiments referred to in chapter IV (pp. 105–113); similar conditions with chemical stimulation were found to exist by Seguin (1956).

A study by Magladery *et al.* (1952) and Teasdall *et al.* (1952) of the low-threshold monosynaptic reflexes in man revealed that chronic upper motor neurone damage goes hand in hand with an enhancement of these reflexes when hemispheral and upper brain-stem lesions are present and is particularly prominent with lower brain-stem or spinal cord involvement. In an analysis in cats with chronic semisections of the spinal cord Teasdall, Magladery and Ramey (1958) brought forth electrophysiological evidence showing a lowered threshold and greater amplitude for polysynaptic reflex discharges which persisted after a high transection of the spinal cord. The early (monosynaptic) responses were less affected by the semisection and with strong "nerve shocks" an apparent reversal took place, the discharge being greater on the control side (Fig. 43).

Thus, a considerable degree of recovery of function in the spinal neurones may take place not only after section of the reticulo- and vestibulo-spinal tracts, but also after severance of the cerebral projections, provided that sufficient time after the operation is allowed. In the first case, the sensitization is probably due predominantly to the prolonged isolation of the spinal neurones from the reticulo- and vestibulo-spinal influences. In the second, the supersensitivity may be caused by the severance of direct cortico-spinal connections, or it may depend on a penultimate sensitization which is the result of a readjustment of the activity of the reticular and vestibular mechanisms. Changes in the influx from the bulbar centres have been shown to play a prominent part in the development of spasticity after cerebral ablations by Magoun and Rhines (1947), Magoun (1950) and his co-workers (Lindsley, 1952). The occurrence of penultimate sensitization in the striated muscle was demonstrated by Cannon and Haimovici (1939) in experiments with semisection of the spinal cord, and in the nictitating membrane of the cat after section of the cervical sympathetic trunk by Hampel (1935). In both these instances peripheral structures were involved but sensitization of the spinal neurones by deafferentation carried out peripherally to the dorsal root ganglion (Luco and Eyzaguirre, 1952) falls into the category of penultimate reactions within the central nervous system.

The relative merits of the release or disinhibition hypothesis compared with those of an augmented influx to lower centres following cerebral ablations were discussed by Magoun and Rhines (1947) in relation to spasticity. Undoubtedly both these factors play their part in the changes which result from damage to the higher levels of the brain. However, both disinhibition and augmented influx from different regions of the central nervous system are sudden alterations which depend upon an anatomical defect and once produced should not be subject to further readjustment. In spite of that, a progressive change in sensitivity to chemical agents and a very gradual increase in complexity of reflex patterns may be observed after a cerebral ablation as well as a gradual recovery from the so-called "shock" of spinal transection. Furthermore, greater sensitivity of the isolated neurones manifests itself not only in the responses to chemical stimulating agents or to reflex excitation but also to depressant drugs and to adverse conditions. Thus pentobarbital, ether, curare, cold, haemorrhage, fatigue and disease affect more the side of the spinal cord which corresponds to the preceding cerebral ablation than the control side. A suggestion that this is so in human beings as well as in experimental animals was made by Walsh (1957, p. 88) who pointed out that restoration of function in patients suffering from spinal shock is delayed by " 'toxaemia' associated with bed-sore or a urinary tract infection."

Whereas a greater sensitivity of the isolated neurones to chemical and to reflex excitation can result in many experimental situations from disinhibition or from facilitatory effects of an "augmented influx," and if progressive recovery may be attributed to the passing of operative trauma, a greater sensitivity to depressant agents and adverse conditions cannot be accounted for in the same way. This may be regarded as one of the most conclusive pieces of evidence for the existence of a true change in the sensitivity of the denervated nerve cells. Once this is conceded, then it follows that the change in sensitivity is likely to play a part in the process of compensation.

On the whole, the concept suggested by Jackson (1884) in order to account for the compensation which occurs after damage to parts of the central nervous system is supported by experimental evidence. This evidence points in two directions. First, after injury to the descending paths, the lower levels of the central nervous system become more responsive to nerve impulses arriving through the remaining connections and to chemical agents. This change probably depends upon several factors. It may be partly due to a release, partly to an augmented influx, and partly to the sensitization by denervation. This latter change allows for a gradual improvement of responses to occur at a distance from the lesion long after the direct effects of the initial trauma wore off. All the factors together produce an "increased flow of energy in an isolated level" in Jackson's terminology.

The second component which may play a part in the process of functional reorganization is the sensitization of the highest levels of the nervous system which takes place after the interruption of afferent pathways represented by the ascending and commissural connections of the brain. An increased generation of impulses at these levels, which results from the change in sensitivity brought about by partial deafferentation, may account for the "increased flow of energy over the remaining connections" to the lower levels, thus opening up the vicarious pathways envisaged by Jackson (1884). Together, the two effects may lead to a partial restitution of function and to the appearance of abnormal responses (pathological signs) which signify, physiologically, an exaggeration and distortion of the attempts of nature at compensation which must occur in the absence of regeneration of nervous tissue.

2.   DERANGEMENTS OF THE EXTRAPYRAMIDAL SYSTEM AND SENSITIZATION BY DENERVATION

An interesting aspect of the exaggerated sensitivity of isolated regions of the central nervous system was revealed in studies of the extrapyramidal system. As pointed out by Stavraky (1960), there is an over-all similarity between the results of isolation of spinal motor neurones and signs of paralysis agitans. Interruption of connections with other areas of the nervous system brought about either by severance of descending paths, deafferentation or selective destruction of interneurones, such as was attained in the experiments of Harreveld and Marmont (1939) and Gelfan and Tarlov (1959) by means of asphyxiation of the spinal cord, or a combination of these factors, results in a poverty and inco-ordination of movements, changes in tone and, at the same time, a great increase of irritability in the motor neurones. The latter aspect of isolation is expressed in a lowered threshold to chemical agents and to impulses arriving at these neurones through remaining connections, the irritability reaching a level where spontaneous discharges of the neurones take place. What bearing can these findings have on the symptoms of Parkinsonism?

Ward, McCulloch and Magoun (1948) and Peterson, Magoun, McCulloch and Lindsley (1949) succeeded in producing Parkinsonian-like tremors in monkeys by bilateral electrolytic lesions in the reticular formation at the base of the tegmentum. Realizing that the site of a destructive lesion cannot initiate a motor response, Ward and Jenkner (1953) and Jenkner and Ward (1953) investigated the possible mechanism of the derangement and have shown that the tremor originates in the medial reticular formation of the brain stem. Ward and Jenkner advanced the hypothesis that in lesions of the rostral tegmentum or lower diencephalon the tremor results from a sensitization of the isolated neurones in the medial reticular formation. Because the tremor was reduced or abolished by anticholinergic drugs, Ward and Jenkner (1953) and Jenkner and Ward (1953) suggested that a Parkinsonian tremor is the result of a supersensitivity of the isolated medial reticular formation to locally liberated acetylcholine.

Support for this hypothesis may be found in the fact that many supraspinal structures seem to be exceedingly sensitive to chemical agents. This was shown by Gesell and his co-workers (Worzniak and Gesell, 1938; Hansen, Worzniak and Gesell, 1942; Gesell and Hansen, 1943) with acetylcholine, eserine and atropine for the respiratory centre and by Miller (1943, 1949) for the hypoglossal nucleus. French, Verzeano and Magoun (1953), Arduini and Arduini (1953, 1954), Rinaldi and Himwich (1955 a, b, c), King (1956), King *et al.* (1957) and Killam and Killam (1957) have made similar observations with cholinergic drugs, barbiturates and other substances in the brain-stem activating system.

On the other hand, studying by means of a microelectrode technique unit discharges of the sensory motor cortex in monkeys with postural tremor induced by brain-stem lesions, Cordeau *et al.* (1960) and Cordeau (1960) concluded that a spontaneously oscillating mechanism is set in motion in telencephalic and diencephalic structures involving the motor cortex, lenticular nucleus (or corpus striatum as a whole) and the nucleus ventralis of the thalamus. According to Cordeau, "the oscillations may be the consequence of a partial deafferentation of these structures and would be under the regulatory influence of a servomechanism provided by the peripheral gamma loop acting through spino-cerebellar and cerebellothalamic pathways." Tremorogenic impulses set up in this central pacemaker would, according to this view, travel down to the anterior horn cells of the spinal cord through pyramidal tract fibres, a concept postulated for clinical Parkinsonian-type tremors (Bucy, 1949, pp. 365–408; 1958, pp. 271–293). Discussing these observations, Walker (1960) and Spiegel (1960) suggested that in hyperkinesis of extrapyramidal origin there may be a dual interplay of centrifugal influence upon the spinal neurones deriving from the cerebral cortex and from the extrapyramidal system itself. Spiegel, in particular, drew attention to the fact that in man destruction of the pyramidal system does not always abolish Parkinsonian tremors and that placement of discrete lesions in the reticular formation is sometimes capable of alleviating this condition. Spiegel *et al.* (1959) also found that following experimental elimination of the head of the caudate nucleus, pallidal electrical discharges are initially reduced but recover after variable intervals and even exceed the preoperative discharges in amplitude. A hyperreactivity to thalamic stimulation and to chemical agents was also observed at that stage. Spiegel *et al.* pointed out that the view that the pallidum receives inhibitory connections from the striatum cannot be maintained and concluded that pallidal hyperactivity following striatal lesions has to be interpreted as an "isolation phenomenon."

It is interesting that, long before Spiegel's studies, Hunt (1917, 1918, 1923, 1931) concluded, largely on the ground of pathological investigations, that degeneration of the small cells in the corpus striatum releases the large motor nerve cells in this structure from control and leads to chorea, whereas additional destruction of large neurones due to diffuse lesions in the globus pallidus results in a Parkinsonian syndrome. From similar observations in other regions of the central nervous system, Hunt suggested that all central inhibition may be the result of the influence of Type II Golgi cells, a view which, in the case of the spinal cord, was considered repeatedly in one way or another in subsequent investigations (Gasser, 1937; Harreveld and Marmont, 1939; Harreveld, 1940, 1943, 1944; Kabat and Grenell, 1944; Brooks and Eccles, 1947; Gelfan and Tarlov, 1959; and others). Although some of Eccles' (1953, pp. 166–73) later work with an intracellular technique of recording nerve cell potentials does not in his own estimation support the Golgi cell hypothesis, Eccles (1955) and Eccles and his co-workers (1954) suggested that the inhibitory action of impulses conducted antidromically in motor nerve fibres occurs through the agency of specific cholinergic internuncials, the so-called "Renshaw Cells" which are activated through axon collaterals and exert their inhibitory effects by way of synaptic connections with motor neurones of the same level. The "Renshaw Cells" are considered by Eccles *et al.* (1954) to exert similar inhibitory effects when activated by impulses conducted orthodromically in the axon of the motor neurone; thus, they may form a basis for a self-limiting backfeed in the spinal centres—a possibility not altogether unlike the one considered by Hunt (1931).

3.   BACKGROUND AND THEORIES OF SUPERSENSITIVITY

Throughout this discussion of sensitivity of partially isolated neurones of the central nervous system a conscious attempt has been made to use evidence brought forth with methods based on observations of the behaviour of animals in chronic experiments or on recordings obtained with the minimum of dissection of the nervous system. Great emphasis is being placed of late on the use of analytical electrophysiological techniques which allow interpretation of the activity of single muscle fibres, nerve cells or peripheral receptors. However, there is a danger that in the process of reducing the preparation to the point at which such observations can be made, the delicate state of sensitivity of the individual units may be altered, and wrong impressions as to their responses may be drawn. Support for this view is found in the fact that operative procedures and exposure of the central nervous system to cold, bleeding, fatigue, curarization and anaesthetics as well as to other adverse conditions lead to a more pronounced depression of the supersensitive areas of the nervous system. Similarly, in animals with long-term lesions or distemper, insufficient time for recovery from operation or from induced convulsions and the use of various drugs temporarily reverses the response, making the control side appear to be more sensitive and more active than the one innervated by partially isolated neurones. In view of these findings, it seemed advisable to rely, for the most part, on indirect evidence obtained under optimal conditions of sensitivity to establish the basic facts about the responses and the behaviour of partially isolated regions of the nervous system in preference to more accurate procedures, which would allow for an analysis and elucidation of the fundamental processes involved in this state.

Some suggestions regarding the possible nature of the phenomena which underlie the changes in irritability following denervation were made by Cannon and Rosenblueth (1949, pp. 187–203). They fall into two main groups: one postulates alterations in permeability and an easier access of chemical agents into the cells, whereas the other deals with cellular metabolism. Into this latter category can be included changes in the activity of cholinesterase and amine oxidase as well as of other specifically active substances and metabolites. Recently, interesting studies pertaining to changes in electrical properties of muscle cell membranes have been reviewed by Thesleff (1960), but they have not advanced to a stage where their bearing on supersensitivity can be assessed in a general way.

An increased rate of penetration of radioactive potassium and phosphorus into rats' muscle following denervation was reported by Noonan, Fenn and Haege (1941) and by Friedlander, Perlman and Chaikoff (1941), and the augmented uptake of potassium in that animal was shown by Lyman (1942) not to depend on circulatory changes. On the other hand, the uptake of radioactive potassium by denervated frog muscle was reduced according to Harris and Nicholls (1956). An increase in membrane resistance thus brought about could enhance the sensitivity to electrical (current) stimulation reported by Nicholls (1956) but would not *per se* change the voltage sensitivity. Also, it could not account for the far greater sensitivity to acetylcholine. Smith (1941) and Parker (1942) found that adrenaline in suitable concentrations causes a more profound and more rapid concentration of pigment in the denervated melanophores of the tautog and catfish, and the observations of Chang, Hsieh and Lu (1939) with sympathetic stimulation in the snake fish gave similar results.

An attempt to analyse the effect of denervation on permeability was made by Wiedeman (1954, 1955) who studied the sensitivity of the vascular bed in the sympathectomized bat's wing to adrenaline. It was found that while intravenous injections were more effective in the sympathectomized wing, the thresholds for local application of this agent varied greatly and were not significantly different in control and denervated wings. From this, the conclusion was drawn that the permeability of the vessel wall is affected by denervation rather than the intrinsic responsiveness of the contractile tissue itself. The fact that sympathectomy renders capillaries in striated muscle more permeable to colloidal dyes was shown by Gabbe (1926) and makes this possibility in the bat's wing quite plausible. The use by Wiedeman (1954, 1955) of successive hourly applications of adrenaline, however, obscures the results as far as the irritability of the contractile elements is concerned, since it was shown by Simeone (1938), Stavraky (1947) and others that the sensitivity of denervated structures to adrenaline declines sharply with repeated applications of this agent. However, this work is important inasmuch as it draws attention to the state of the circulation in the denervated area. It is known that partially isolated regions of the central nervous system acquire a greater vascularity than the surrounding nervous tissue and this may play a part in the responses of the denervated regions to blood-borne agents, though it would not explain the greater neuronal excitability to impulses arriving by way of the remaining nervous connections. Also Echlin (1959) in an electroencephalographic analysis of the supersensitivity of denervated cortical neurones excluded the influence of changes in vascularity by special refinements in experimental technique.

As suggested by Welsh (1955, p. 133), an uncovering of the receptors or synaptic surfaces following degeneration of the nervous terminations resulting in greater accessibility of the cells to the chemical agents could account for many of the observed effects. An expansion of the receptive substance to involve the whole muscle fibre is also postulated as the underlying mechanism of supersensitivity to acetylcholine in the case of the denervated muscle by Axelsson and Thesleff (1959) and Miledi (1959), a diffuse sensitivity of foetal muscle to acetylcholine described by Diamond and Miledi (1959) falling into the same category of observations. The attempt of Cannon and Haimovici (1939) to correlate the increase in sensitivity of spinal motor neurones to degenerative changes in terminal buttons after section of the centrifugal fibres in the spinal cord may be regarded as another morphological visualization of this phenomenon. However, these concepts would not explain the penultimate sensitization of a muscle observed by Cannon and Haimovici (1939), Rosenblueth, Lissak and Lanari (1939) and by Solandt and Magladery (1942) on section of descending connections in the spinal cord, sensitization of normally innervated end-plates in the muscle following denervation of other end-plates on the same cell (Frank, 1959), or sensitization of the denervated end-plate region as observed by Kuffler (1943). Similarly, it would not account for sensitization of the nictitating membrane by decentralization (Hampel, 1935), or for an increased sensitivity of spinal motor neurones seen by Luco and Eyzaguirre (1952) after severance of the dorsal nerve roots peripherally to their ganglia.

As seen by Drake and Stavraky (1948 b), the onset of supersensitivity of spinal neurones after deafferentation did not develop to all the tested agents simultaneously; the neurones became more sensitive first to physiological stimulants, such as acetylcholine and adrenaline, and only later to strychnine, camphor and pentylenetetrazol.

Thus on the whole, as conceded by Cannon and Rosenblueth (1949, pp. 202–3) and concurred in by many subsequent investigators, changes in permeability cannot account for all the phenomena of supersensitivity, and other factors must be involved in this condition. One possible explanation is sought in changes in acetylcholine metabolism and in cholinesterase activity. As far as the submaxillary gland is concerned, it is noteworthy that the greater sensitivity to acetylcholine and the prolonged response to it following denervation go hand in hand with a reduction in the maximal rate of secretion (Graham and Stavraky, 1954 c). This would be in keeping with the belief that a lowering of the cholinesterase activity in the denervated gland is responsible for the change. Recently, more direct evidence in support of such a view has been brought forth by Stromblad (1955) who found that parasympathetic denervation of the salivary glands caused a thirty to sixty per cent decline in the amount of cholinesterase in them, these values depending respectively on whether the denervation was pre- or post-ganglionic. However, in order to establish the relation between this interesting finding and the supersensitivity of the denervated gland to acetylcholine, further study is required, particularly in view of the fact that previous investigations failed to detect any reduction of cholinesterase activity under approximately similar experimental conditions (Beznak, 1932; Chang and Gaddum, 1933; MacIntosh, 1937).

Changes in cholinesterase activity were reported in other denervated tissues. The great increase in sensitivity to acetylcholine of denervated muscle is well known (Brown, 1937; Rosenblueth and Luco, 1937; Solandt and Magladery, 1942; Kuffler, 1943; Nicholls, 1956 and others) and Feng and Ting (1938), Snell and McIntyre (1955) and Beckett and Bourne (1957) described a decrease of cholinesterase in the end-plates of striated muscles after section and degeneration of the motor nerves. Brücke (1937) and Couteaux and Nachmansohn (1940) found the same to be true in the superior cervical sympathetic ganglion after section of pre-ganglionic fibres. Martini and Torda (1938) observed a reduction of the cholinesterase in the spinal cord below a transection, while Nachmansohn and Hoff (1944) noted a similar condition after section and degeneration of the dorsal roots. Marrazzi *et al.* (1955, 1956) summarized the recent evidence which suggests that transmission phenomena at central synapses are influenced in an excitatory manner by cholinergic drugs. However, the role of acetylcholine in many instances of central transmission is questioned even by advocates of the hypothesis of chemical mediation (Feldberg *et al.*, 1951; Hebb, 1957), and the part played by this agent in supersensitivity caused by partial isolation of various regions of the central nervous system cannot be precisely assessed at this time.

It was suggested by Blaschko, Richter and Schlossmann (1937) that adrenaline is inactivated in vertebrate organisms by amine oxidase. In accordance with this Burn and Robinson (1952, 1953) found that the increased sensitivity of the denervated nictitating membrane of the cat to adrenaline and noradrenaline, and the prolongation of the responses brought about by these agents was actually associated with a decrease in the amine oxidase content of the nictitating membrane. If a similar reduction of amine oxidase should be found in other adrenergic effectors after denervation, and particularly in some central synapses, this could account for certain instances of supersensitivity. According to Burn and Rand (1958) the fact that post-ganglionic denervation of involuntary muscles sensitizes them to the action of adrenaline and noradrenaline (Meltzer and Auer, 1904; Burn and Hutcheon, 1949;

Armin, Grant, Thompson and Tickner, 1953) could also be accounted for by a diminution of the noradrenaline content of the denervated tissues. A decline in the noradrenaline content of various organs following section and degeneration of sympathetic fibres was shown to take place by these authors as well as by von Euler and Purkhold (1951) and by Goodall (1951). However, sensitization by denervation is not specific: excision of the chorda tympani or of the superior cervical sympathetic ganglion increases the sensitivity of the submaxillary gland to both acetylcholine and adrenaline as well as to other drugs (Graham and Stavraky, 1954 c, Emmelin, 1960); whereas isolation of areas of the central nervous system sensitizes the neurones to a variety of artificial stimulating and depressing agents which bear little relation chemically to each other or to cholinergic and adrenergic agents.

The possible decrease of an inhibitory substance in denervated regions of the central nervous system should not be overlooked. As referred to before, Florey (1953, 1954 a, b, 1956), Florey and McLennan (1955 a, b, c), Elliott and Florey (1956), Pataky and Pfeifer (1955) and Pfeifer and Pataky (1955) report on the chemical and physiological characteristics of such factors, but no direct evidence of their role in supersensitivity of denervated structures is available at this time.

Changes in irritability may also occur in the absence of damage to nervous connections. As mentioned previously (chapter III, p. 73, and chapter IV, p. 103), proper spacing of stimuli and other conditions involving activation of individual units may lead to facilitation and augmented excitability of a tissue as well as to a refractory state in it. Discussing changes of state not visible in potential such as facilitation and other after effects of stimulation, Bullock (1959) remarks: "Whereas the classical concept of the neurone recognized the importance of excitability, this has been measured or thought of in terms of the spike threshold. . . . What the newer knowledge has added is understanding that prior to initiation of the spike there are critical forms of excitability not measured by thresholds, because they determine the responses of subthreshold, graded and local events as a function of what came before them." Long before modern refinements in technique made possible this conclusion in regard to states of augmented excitability, it was shown that immediately after stimulation living tissues become partly or completely inexcitable. This refractory period is supposed to be due to a breakdown and subsequent restoration of specific chemical compounds or to their displacement and regrouping. Furthermore, a so-called "receptor desensitization" in normal tissues may be brought about by acetylcholine (Brown, Dale and Feldberg, 1936) and by repetitive stimulation of motor nerves (Wedensky, 1903). A review of the latest work pertaining to these occurrences in the striated muscle is given by Thesleff (1960). A similar process can conceivably take place after denervation. Wolff and Cattell (1937) sensitized the nictitating membrane in cats by pre-ganglionic section of the sympathetic nerve and found it to be three times more responsive to adrenaline. Tetanic stimulation of the post-ganglionic sympathetic fibres markedly reduced this sensitivity to adrenaline; after cessation of stimulation, the sensitivity returned over a period of twenty minutes to the pre-stimulatory level. Simeone (1938) noted that a more lasting depression could be brought about by adrenaline. When 2 mgm. of adrenaline were injected subcutaneously three times a day for five days, the contractions of the denervated nictitating membrane became reduced, while the intact membrane on the opposite side was sensitized by this procedure. In the central nervous system Stavraky (1943, 1947) encountered similar conditions: frontal-lobectomized or semidecerebrated cats showed a selective sensitivity of the isolated regions of the nervous system to acetylcholine and adrenaline as long as the intravenous injections were spaced one week or longer apart. If injected more often, these drugs depressed the denervated neurones, a reversal of the motor responses taking place on the two sides of the body. Drake, Seguin and Stavraky (1956) confirmed these observations and noted that in the isolated quadriceps muscle protracted effects of injections of strychnine would start off with an earlier and much greater response on the sensitized side but after some time the contractions of the decentralized quadriceps became weaker than those of the control. Some of these considerations led Loewi (1949) to suggest that the supersensitivity may be due to inactivity and accumulation of the metabolites of rest in the denervated regions. It is interesting that prolonged administration of atropine and other parasympatholytic agents sensitizes the submaxillary gland of the cat to acetylcholine and to adrenaline (Emmelin, Jacobson and Muren, 1951; Emmelin, 1960) and that a similar administration of dibenzyline increases the sensitivity of the nictitating membrane to adrenaline and noradrenaline (Nickerson and House, 1958). Immobilized muscles with intact nerve supply have also been reported to become hypersensitive to acetylcholine (Solandt, Partridge and Hunter, 1942). However, after tenotomy Desmedt (1949, 1959) found no increase in excitability of the anterior tibial muscle of the rabbit as judged by its latency, chronaxie, voltage capacity curve, anodic opening contraction and other properties. Furthermore, as stated before, a greater sensitivity was found to develop after deafferentation when the dorsal roots were severed centrally to the ganglion and a lesser one when they were cut peripherally (Luco and Eyzaguirre, 1952). The alteration in tone was apparently the same in both instances and changes in activity could not account for this difference. Similarly, after cerebral ablations Stavraky (1943, 1947) and Drake, Seguin and Stavraky (1956) observed an increased tone of the corresponding extremities which was associated with good use of the limbs in walking and stance. The sum total of activity in such limbs was probably greater than in the intact extremities, yet acetylcholine, adrenaline and convulsant drugs caused exaggerated responses in decentralized limbs.

As mentioned previously, Thesleff (1960) suggests that due to the electric core conductor properties an increase in the chemically sensitive surface of a denervated muscle fibre renders it supersensitive to depolarizing agents. It is interesting that a study of electrical potentials in denervated muscle led Li (1960) to the conclusion that spontaneous fibrillation potentials do not originate in specific focal areas in muscle fibres, but are due to differences in membrane properties of these fibres which are probably associated with changes in metabolic activity. After denervation, an increase in hexokinase and a decrease in cytochrome oxidase activity were observed in the muscle by Humoller, Hatch, and McIntyre (1951, 1952). These changes seem to go hand in hand with a decrease in resting membrane potential and an increase in the oscillation of the potential. The "oscillating potentials" differ from the "miniature end-plate potentials" observed in normally innervated muscle fibres by Fatt and Katz (1951, 1952) in that they are not all-or-none and random, but are graded and rhythmic. Li believes that the decrease and instability of resting potentials may lead to rhythmic spike discharges in the fibrillating muscle fibres and that a similar process occurs in pathological nerve cells of the cerebral cortex during epileptiform activity.

So far, the general properties of living tissues which may influence their excitability have been considered. However, the complicated anatomical arrangement of the central nervous system presents additional factors which may play a part in supersensitivity caused by partial denervation. In the first section of this chapter the relative roles of "disinhibition" and "augmented influx" versus "sensitization by denervation" have been mentioned in relation to the restoration of function after anatomical damage to the central nervous system. Lorente de No (1939) has invoked another principle: that of reverberating circuits to explain states of overactivity which develop in cranial motor nuclei. Accordingly, one neurone activates the second, this a third, and so on until the last one reactivates the initial neurone. This results in a series of trapped impulses running round and round in neurone circles. The concept resembles that employed by Mines (1913), Garrey (1924) and Lewis (1925) to explain the origin of the repetitive excitation in the heart which occurs in auricular fibrillation.* Similarly, reverberating neuronal nets could maintain abnormal dynamic states in partially isolated areas of the central nervous system, and this theory was widely applied in clinical neurology to account for hyperkinetic states such as occur in extrapyramidal disease (Bucy, 1949, pp. 365–408), and was also considered in causalgia (Ash, 1952), while Burns (1951, 1954) and Halpern and Sharpless (1959) showed that analogous conditions prevail in isolated slabs of cerebral cortex in which excitation leads to prolonged after-discharges. The prolonged state of facilitation which follows the excitation of the central end of a previously severed dorsal root (Eccles and McIntyre, 1951, 1953), and the finding of Drake and Stavraky (1948 b) that actual fusion of individual contractions of the quadriceps takes place on the deafferented side while a series of discrete patellar tendon jerks are evoked in the control extremity, fit in well with Burns' observations.

Another possible explanation of these findings was suggested by Gerard (1951) when considering the mechanism of causalgia. "In the cord," he writes, "a hypersynchronization, a firmer locking together of a larger than normal number of neurones has occurred, to form a pulsating pool, and this synchronization has become exaggerated by virtue of the lack of disturbing impulses to disrupt the synchrony and by reinforcement with those specific pain afferents that are feeding in to lock the neurones. Such a pulsating pool could recruit additional units, could move along in the grey matter, could be maintained by impulses different from and feebler than those needed to initiate it, could discharge excessive and abnormally patterned volleys."

Finally, "sprouting" of the terminations of nerve fibres in response to denervation of adjacent areas has been described by Weddell *et al.* (1941), Hoffman (1950), Edds (1950) and by Murray and Thompson (1957). Recently, McCouch *et al.* (1958) found this condition to exist in the spinal cord after aseptic semisection of the latter. In this instance sprouting became evident after an interval of two to three weeks while reflex responses became exaggerated on the side of the semisection considerably earlier. This led McCouch *et al.* to conclude that other factors must be involved in the production of spasticity as well as sprout-

*This mechanism of auricular fibrillation has been questioned recently by Burn. (Burn, J. H. The cause of fibrillation. Can. Med. Assoc. J. *84:* 625–627)

ing; the authors failed to find evidence of an increase in internuncial potentials greater than that from afferent terminals, and seemed to feel that this finding is contrary to the concept of supersensitivity. It is difficult to see how this observation could speak either for or against a concept which is based primarily on the results of chemical stimulation as well as on other evidence which indicates a lowered threshold of isolated regions of the nervous system without predetermining its seat of origin. On the other hand, sprouting itself may be an important morphologic component of the later phases of functional compensation in the nervous system.

Both the theory of reverberating circuits and that of hypersynchronization as well as the phenomenon of sprouting and other factors could account for many features of centrally occurring supersensitivity. However, it must be remembered that there exists a peculiar affinity of denervated structures to chemical agents both stimulant and depressant. This affinity is very pronounced towards narcotics and particularly towards barbiturates in the central nervous system (Seguin and Stavraky, 1957). These agents depress denervated regions of the central nervous system in very weak concentrations, and their effect can hardly be accounted for by an interference with reverberating circuits or with a hypersynchronized activity of denervated neuronal pools or other known factors because it results in a selective loss of function of the areas in question and not in a general suppression of activity. Similar circumstances prevail in sympathetic ganglia and in peripheral effector cells where intercalated neurones do not exist.

A specific process seems to be involved in supersensitivity which is related to the basic property of excitability of living matter and until this latter is more fully understood the true nature of supersensitivity will probably remain obscure. A quotation from Herrick (1956, p. 458), in which he adopts a Newtonian approach, may be appropriate here. Discussing psi phenomena in relation to the dimensions of the mind, he writes: "we do not regard gravity as a spurious phenomenon because we do not know how to explain it. What we try to do is to find out how it works, and if no current theories fit the facts we search for other principles that satisfy the requirements." In the present state of our knowledge of many biological phenomena this attitude could justifiably be applied to supersensitivity.

REFERENCES

ABDON, N. O.
1945. On metabolism of acetylcholine precursor in isolated hearts. Acta Pharmacol. Toxicol. *1*: 169–83.

ADRIAN, E. D.
1928. The basis of sensation: the action of sense organs. Christophers, London. 122 pp.
1931. The messages in sensory nerve fibres and their interpretation. Croonian Lecture. Proc. Roy. Soc. (London). B *109*: 1–18.
1932. The mechanism of nervous action: electrical studies of the neurone. Oxford University Press, London, 103 pp.
1941. Afferent discharges to the cerebral cortex from peripheral sense organs. J. Physiol. *100*: 159–191.
1953. Sensory messages and sensation: the response of the olfactory organ to different smells. Acta Physiol. Scand. *29*: 5–14.
1955. Synchronized activity in the vomero-nasal nerves with a note on the function of the Organ of Jacobsen. Pflüger's Arch. ges. Physiol. *260*: 188–192.

ADRIAN, E. D., and BRONK, D. W.
1929. The discharge of impulses in motor nerve fibres. Part II. The frequency of discharge in reflex and voluntary contraction. J. Physiol. *67*: 119–159.

ADRIAN, E. D., and UMRATH, K.
1929. The impulse discharge from the Pacinian corpuscle. J. Physiol. *68*: 139–154.

ADRIAN, E. D., and ZOTTERMAN, Y.
1926a. The impulses produced by sensory nerve endings. Part 2. The response of a single end-organ. J. Physiol. *61*: 151–171.
1926b. The impulses produced by sensory nerve endings. Part 3. Impulses set up by touch and pressure. J. Physiol. *61*: 465–483.

AKELAITIS, A. J., RISTEEN, W. A., HERREN, R. Y., and VAN WAGENEN, W. P.
1942. Studies on corpus callosum: contribution to study of dyspraxia and apraxia following partial and complete section of corpus callosum. Arch. Neurol. Psychiat. *47*: 971–1008.

ALCOCK, P., BERRY, J. L., and DALY DE BURGH, I.
1935. Action of drugs and pulmonary circulation. Quart. J. Exp. Physiol. *25*: 369–392.

ALTAMIRANO, M., FERNANDEZ, E., and LUCO, J. V.
1949. Comparative study of antagonistic drugs on normal and denervated effects. Am. J. Physiol. *156*: 280–284.

142 *References*

ANOCHIN, P., and ANOCHINA-IVANOVA, A.
  1929. Ueber die vasomotorische und sekretorische Reaktion der Speicheldrüse auf Einführung von Acetylcholin. Pflüger's Arch. ges. Physiol. *222*: 478–492.

ANREP, G. V., and SEGALL, H. N.
  1926. Regulation of coronary circulation. Heart *13*: 239–260.

ARDUINI, A., and ARDUINI, M. G.
  1953. Action of drugs and metabolic alterations on brain stem activating system. Federation Proc. *12*: 6.
  1954. Effect of drugs and metabolic alterations on brain stem arousal mechanism. J. Pharmacol. Exp. Therap. *110*: 76–85.

ARMIN, J., GRANT, R. T., THOMPSON, R. H. S., and TICKNER, A.
  1953. Explanation for heightened vascular reactivity of denervated rabbit's ear. J. Physiol. *121*: 603–622.

ARMITAGE, G., and MEAGHER, R.
  1933. Gliomas of corpus callosum. Z. ges. Neurol. Psychiat. *146*: 454–488.

ASH, W. H.
  1952. Pain Mechanisms. U.S. Armed Forces Med. J. *3*: 1257–1266.

ASRATIAN, E. A.
  1953. Fisiologia centralnoi nervnoi sistemi. Press of the Academy of Science, Moscow, U.S.S.R., 559 pp.

ASUAD, J.
  1940. Epilepsie expérimentale. Presse méd. *48*: 794–798.

AXELSSON, J., and THESLEFF, S.
  1959. A study of supersensitivity in denervated mammalian skeletal muscle. J. Physiol. *147*: 178–193.

BABKIN, B. P.
  1913. Die Arbeit der Speicheldrüsen beim Hunde nach Entfernung des Ganglion cercicale superior sympathici. Pflüger's Arch. ges. Physiol. *149*: 521–531.
  1946. Antagonistic and synergistic phenomena in the autonomic nervous system. Trans. Roy. Soc. Can., 3rd series, Sect. V, *40*: 1–25.

BABKIN, B. P., ALLEY, A., and STAVRAKY, G. W.
  1932. Humoral transmission of chorda tympani effect. Trans. Roy. Soc. Can., 3rd series, Sect. V, *26*: 89–107.

BABKIN, B. P., GIBBS, O. S., and WOLFF, H. G.
  1932. Die humorale Uebertragung der Chorda tympani-Reizung. Arch. exp. Path. Pharmakol. *168*: 32–37.

BAGLIONI, S.
1900. Physiologische Differenzirung verschiedener Mechanismen des Rückenmarkes. Arch. Anat. u. Physiol., Supplement B, pp. 193–242.

BAIN, W. A.
1932. On the mode of action of vasomotor nerves. J. Physiol. *77*: 3–4P.

BARANY, E. H., and STEIN-JENSEN, E.
1946. The mode of action of anticonvulsant drugs on electrically induced convulsions in the rabbit. Arch. intern. pharmacodynamie *73*: 1–47.

BARD, P.
1933. Studies on the cerebral cortex. Arch. Neurol. Psychiat. *30*: 40–74.

BARRON, D. H., and MATTHEWS, B. H. C.
1935. Intermittent conduction in the spinal cord. J. Physiol. *85*: 73–103.
1938. The interpretation of potential changes in the spinal cord. J. Physiol. *92*: 276–321.

BAYLISS, W. M.
1908. Note on the supposed existence of vaso-constrictor fibres in the chorda tympani nerve. J. Physiol. *37*: 256–263.

BECHTEREW, W.
1883. Ergebnisse der Durchschneidung des Nerwen acusticus nebst Erörterung der Bedeutung der semicirculären Canäle für das Korpergleichgewicht. Pflüger's Arch. ges. Physiol. *30*: 312–347.

BECKETT, E. B., and BOURNE, G. H.
1957. Histochemical demonstration of cholinesterase and 5-nucleotidase in normal and diseased human muscle. Nature *179*: 771–772.

BELING, C. C., and MARTLAND, H. S.
1919. A case of tumor of the corpus callosum and frontal lobes. J. Nervous Mental Disease *50*: 425–432.

BENDER, M. B.
1938. Fright and drug contractions in denervated facial and ocular muscles of monkeys. Am. J. Physiol. *121*: 609–619.

BERITOFF, J. S.
1915. On the mode of origination of labyrinthine and cervical tonic reflexes and on their part in the reflex reactions of the decerebrate preparation. Quart. J. Exp. Physiol. *9*: 199–229.

BERNARD, C.
1880. Leçons de pathologie expérimentale. 2nd ed. Baillière, Paris. 604 pp.

BERNHARD, C. G., and SKOGLUND, C. R.
1942. Selective activation of a transient reflex by restricting stimulation to certain frequencies. Acta Physiol. Scand. *4*: 125–135.
1953. Potential changes in spinal cord following intra-arterial administration of adrenaline and noradrenaline as compared with acetylcholine effects. Acta Physiol. Scand. *29*, Suppl. 106: 435–454.

BERTHA, H.
1928. Zur Abhängigkeit des klonischen Krampfes von der Hirnrinde. Z. ges. exp. Med. *58*: 187–195.

BEZNAK, A. V.
1932. Ueber die autokoide Aktivität der Chorda tympani. Pflüger's Arch. ges. Physiol. *231*: 400–409.

BEZNAK, M., and FARKAS, E.
1937. Interpretation of some phenomena of salivary secretion caused by direct electrical stimulation of effector nerve, in terms of present knowledge of nervous impulse and of its chemical transmission. Quart. J. Exp. Physiol. *26*: 265–283.

BIGNAMI, A., and NAZARI, A.
1915. Sulla degenerazione delle commissure encefaliche e degli emisferi nell' alcoolismo cronico. Riv: sper. freniat. *41*: 81–148.

BISHOP, G. H., and LANDAU, W. M.
1958. Evidence for double peripheral pathway for pain. Science *128*: 712–713.

BLASCHKO, H., RICHTER, D., and SCHLOSSMANN, H.
1937. The inactivation of adrenaline. J. Physiol. *90*: 1–17.

BOHROD, M. G.
1942. Primary degeneration of corpus callosum (Marchiafava's disease): report of second American case. Arch. Neurol. Psychiat. *47*: 465–473.

BONNET, V., and BREMER, F.
1937a. Action du potassium, du calcium et de l'acetylcholine sur les activités électriques spontanées et provoquées de l'écorce cérébrale. Compt. rend. soc. biol. *126*: 1271–1275.
1937b. A study of the after-discharge of spinal reflexes of the frog and toad. J. Physiol. *90*: 45–47P.

BONVALLET, M., DELL, P., and HIEBEL, G.
1954. Tonus sympathique et activité électrique corticale. Electroencephalog. and Clin. Neurophysiol. *6*: 119–144.

BORK, J. J. VAN, DALDERUP, L. M., and HIRSCHEL, M. I.
1950. Convulsions chez les chiens adultes et non-adultes après administration de cardiazol. Acta Physiol. et Pharmacol. Neerl. *1*: 533–539.

BRADLEY, K., EASTON, D. M., and ECCLES, J. C.
1953. An investigation of primary or direct inhibition. J. Physiol. *122*: 474–488.

BRADLEY, K., and ECCLES, J. C.
1953. Strychnine as depressant of primary inhibition. Nature *171*: 1061–1062.

BRADLEY, K., and SCHLAPP, W.
1950. Effect of strychnine on spinal reflexes in decapitated cats. J. Physiol. *111*: 62P.

BREMER, F.
1928. De l'exagération des réflexes, consecutives à la section des racines postérieures. Ann. physiol. physicochim. biol. *4*: 750–752.
1929. Nouvelles recherches sur la sommation d'influx nerveux. Compt. rend. soc. biol. *102*: 332–336.
1931. Contribution à l'étude du phénomène de l'inhibition centrale. Compt. rend. soc. biol. *106*: 465–469.
1932. Le tonus musculaire. Ergeb. Physiol. *34*: 678–740.
1938. Effets de la déafférentation complète d'une région de l'écorce cérébrale sur son activité électrique spontanée. Compt rend. soc. biol. *127*: 355–358.
1944. La mode d'action de la strychnine à la lumière de travaux récents. Arch. intern. pharmacodynamie, *69*: 249–264.
1953. Postural mechanisms. XIX Intern. Physiol. Congr., Montreal, pp. 72–76.
1954. Anatomo-physiologie de la douleur. Acta Neurol. Psychiat. Belg. *54*: 771–785.

BREMER, F., and BONNET, V.
1942. Contributions à l'étude de la physiologie générale des centres nerveux. II. L'inhibition réflexe. Arch. intern. physiol. *52*: 153–194.

BREMER, F., BONNET, V., and MOLDAVER, J.
1942. Contributions à l'étude de la physiologie générale des centres nerveux. III. L' "after-discharge" réflexe et la théorie neuro-chimique de l'activation centrale. Arch. intern. physiol. *52*: 215–248.

BREMER, F., BRIHAYE, J., and ANDRÉ-BALISAUX, G.
1956. Physiologie et pathologie du corps calleux. Arch. suisses neurol. et psychiat. *78*: 31–87.

BRONDGEEST, P. J.
1860. Untersuchungen über den Tonus der willkürlichen Muskeln. Arch. Anat., Physiol. u. wissenschaftliche Medicin, pp. 703–704.

BROOKS, C. MC., and ECCLES, J. C.
1947. An electrical hypothesis of central inhibition. Nature *159*: 760–764.

BROOKS, V. B., CURTIS, D. R., and ECCLES, J. C.
1955. Mode of action of tetanus toxin. Nature *175*: 120–121.

BROWN, G. L.
1937. The actions of acetylcholine on denervated mammalian and frog's muscle. J. Physiol. *89*: 438–461.

BROWN, G. L., DALE, H. H., and FELDBERG, W.
1936. Reactions of the normal mammalian muscles to acetylcholine and to eserine. J. Physiol. *87*: 394–424.

BROWN, G. L., and HARVEY, A. M.
1938. Reaction of avian muscle to acetylcholine and eserine. J. Physiol. *94*: 101–117.

BROWN, T. GRAHAM
1911a. Studies in the physiology of the nervous system. VIII. Neural balance and reflex reversal with a note on progression in the decerebrate guinea-pig. Quart. J. Exp. Physiol. *4*: 273–288.
1911b. Studies in the physiology of the nervous system. IX. Reflex terminal phenomena—rebound—rhythmic rebound and movements of progression. Quart. J. Exp. Physiol. *4*: 331–397.
1912. Studies in the physiology of the nervous system. XI. Immediate reflex phenomena in the simple reflex. Quart. J. Exp. Physiol. *5*: 237–307.
1913. On postural and non-postural activities of the mid-brain. Proc. Roy. Soc. (London) *B 87*: 145–163.
1915. On the effect of artificial stimulation of the red nucleus in the anthropoid ape. J. Physiol. *49*: 185–194.

BROWN, T. G., and SHERRINGTON, C. S.
1912: The rule of reflex response in the limb reflexes of the mammal and its exceptions. J. Physiol. *44*: 125–130.

BRÜCKE, F. TH. V.
1937. The cholinesterase in sympathetic ganglia. J. Physiol. *89*: 429–437.

BUCY, P. C.
1949. The precentral motor cortex. 2nd ed., Univ. of Illinois Press, Urbana, Ill. 615 pp.
1958. The corticospinal tract and tremor. *In* Houston Neurological Society: Pathogenesis and treatment of Parkinsonism, ed. W. S. FIELDS, C. C Thomas, Springfield, Ill. 372 pp.

BÜLBRING, E.
1944. The action of adrenaline on transmission in the superior cervical ganglion. J. Physiol. *103*: 55–67.

BÜLBRING, E., and BURN, J. H.
1934. The sympathetic dilator fibres in the muscle of the cat and dog. J. Physiol. *83*: 483–501.

1936. The Sherrington phenomenon. J. Physiol. *86*: 61–76.
1941. Observations bearing on synaptic transmission by acetylcholine in spinal cord. J. Physiol. *100*: 337–368.
1942. On action of adrenaline on transmission in sympathetic ganglia which may play a part in shock. J. Physiol. *101*: 289–303.
1949. Action of acetylcholine on rabbit auricles in relation to acetylcholine synthesis. J. Physiol. *108*: 508–524.

BÜLBRING, E., BURN, J. H., and SKOGLUND, C. R.
1948. The action of acetylcholine and adrenaline on the flexor and extensor movements evoked by stimulation of the descending motor tracts. J. Physiol. *107*: 289–299.

BÜLBRING, E., and WHITTERIDGE, D.
1941. Effects of adrenaline on nerve action potentials. J. Physiol. *99*: 201–207.

BULLOCK, T. H.
1959. Neurone doctrine and electrophysiology. Science *129*: 997–1002.

BURGEN, A. S. V.
1956. The secretion of potassium in saliva. J. Physiol. *132*: 20–39.

BURN, J. H.
1945. The relation of adrenaline to acetylcholine in the nervous system. Physiol. Revs. *25*: 377–394.
1950. Relation of motor and inhibitor effects of local hormones. Physiol. Revs. *30*: 77–193.

BURN, J. H., and HUTCHEON, D. E.
1949. Action of noradrenaline. Brit. J. Pharmacol. *4*: 373–380.

BURN, J. H., and RAND, M. J.
1958. The cause of the supersensitivity of organs after degeneration of sympathetic nerves. Lancet *2*: 1213.

BURN, J. H., and ROBINSON, J.
1951. Reversal of vascular response to acetylcholine and adrenaline. Brit. J. Pharmacol. *6*: 110–119.
1952. Effect of denervation on amine oxidase in structures innervated by the sympathetic. Brit. J. Pharmacol. *7*: 304–318.
1953. Hypersensitivity of denervated nictitating membrane and amine oxidase. J. Physiol. *120*: 224–229.

BURN, J. H., and VANE, J. R.
1949. Relation between motor and inhibitor actions of acetylcholine. J. Physiol. *108*: 104–115.

BURNS, B. D.
1951. Some properties of the isolated cerebral cortex in the unanaesthetized cat. J. Physiol. *112*: 156–175.

1954. The production of after-bursts in isolated unanaesthetized cerebral cortex. J. Physiol. *125*: 427–446.
1958. The mammalian cerebral cortex. Edward Arnold, London. 119 pp.

BURTON, A. C.
1951. On physical equilibrium of small blood vessels. Am. J. Physiol. *164*: 319–329.

CAMP, W. J. R.
1928. Pharmacology of cardiazol. J. Pharmacol. Exp. Therap. *33*: 81–92.

CANNON, W. B.
1939. A law of denervation. Am. J. Med. Sci. *198*: 737–750.

CANNON, W. B., and HAIMOVICI, H.
1939. The sensitization of motoneurones by partial "denervation." Am. J. Physiol. *126*: 731–740.

CANNON, W. B., and ROSENBLUETH, A.
1936. Sensitization of sympathetic ganglion by pre-ganglionic denervation. Am. J. Physiol. *116*: 408–413.
1937. Transmission of impulses through sympathetic ganglion. Am. J. Physiol. *119*: 221–235.
1949. The supersensitivity of denervated structures: a law of denervation. Macmillan Co., New York, 245 pp.

CANNON, W. B., ROSENBLUETH, A., and GARCÍA RAMOS, J.
1945. Sensibilización de las neuronas espinales por denervación parcial. Arch. inst. cardiol. Méx. *15*: 327–348.

CARDIN, A.
1946a. Rigidità da decerebrazione negli arti posteriori deafferentati. Boll. soc. ital. biol. sper *22*: 81–83.
1946b. Sul meccanismo della inibizione reciproca degli arti posteriori del l'animale decerebrato. Boll. soc. ital. biol. sper. *22*: 607–608.
1946c. Influenza della deafferentazione di uno o più arti sul tono posturale degli altri arti nell 'animale normale. Boll soc. ital. biol. sper. *22*: 609–610.
1951. Il tono muscolare non è di natura riflessa. Boll. soc. ital. biol. sper. *27*: 1178–1180.
1952a. Il tono muscolare non è natura riflessa. II. Attività del midollo spinale deafferentato. Boll. soc. ital. biol. sper. *28*: 368–371.
1952b. Il tono muscolare non è di natura riflessa. III. Attività del midollo spinale. Boll. soc. ital. biol. sper. *28*: 371–373.

CECIL, R. L.
1944. A textbook of medicine. 6th ed. W. B. Saunders Co., Philadelphia and London. 1526 pp.

CHANG, H. C., and GADDUM, J. H.
1933. Choline esters in tissue extracts. J. Physiol. *79*: 255–285.

CHANG, H. C., HSIEH, W. M., and LU, Y. M.
1939. Light-pituitary reflex and the adrenergic-cholinergic sympathetic nerve in a teleost. Proc. Soc. Exp. Biol. Med. *40*: 455–456.

CHANG, HSIANG-TUNG
1953. Cortical response to activity of callosal neurones. J. Neurophysiol. *16*: 117–131.

CHAVEZ, M., and SPIEGEL, E. A.
1957. The functional state of sensory nuclei following deafferentation. Confinia. Neurol. *17*: 144–152.

CHUSID, J. G., KOPELOFF, L. M., and KOPELOFF, N.
1953. Convulsive threshold values to parenterally injected pentamethylenetetrazol in epileptic monkeys. J. Appl. Physiol. *6*: 139–141.

CLARK, G. A.
1935. Adrenaline vaso-dilatation in voluntary muscle. J. Physiol. *84*: 344–350.

COBB, S.
1924. Electromyographic studies of experimental convulsions. Brain *47*: 70–75.

COMAN, F. D.
1928. Observations on a supposed role of the dorsal roots in muscle tonus. Anat. Record. *38*: 42.

COOK, W. H., and STAVRAKY, G. W.
1952. A cerebellar component of convulsive manifestations. A.M.A. Arch. Neurol. Psychiat. *68*: 741–754.

COOK, W. H., WALKER, J. H., and BARR, M. L.
1951. A cytological study of transneuronal atrophy in cat and rabbit. J. Comp. Neurol. *94*: 267–292.

COOK, L. C., and WALTER, W. G.
1938. Electroencephalogram in convulsions induced by cardiazol. J. Neurol. Psychiat. *1*: 180–186.

COOMBS, H. C.
1932. Effects of repeated electrical stimulation of cortical motor area in cat. Am. J. Physiol. *100*: 64–67.

CORDEAU, J. P.
1960. Microelectrode studies in monkeys with a postural tremor. *In* International symposium on the Extrapyramidal System and Neuroleptics, held at University of Montreal. (In Press)

CORDEAU, J. P., GYBELS, J., JASPER, H., and POIRIER, L. J.
1960. Microelectrode studies of unit discharges in the sensori-motor cortex: investigations in monkeys with experimental tremor. Neulogy *10*: 591–600.

COUTEAUX, R., and NACHMANSOHN, D.
1940. Changes of cholinesterase at end plates of voluntary muscle following section of sciatic nerve. Proc. Soc. Exp. Biol. Med. *43*: 177–181.

CREED, R. S., DENNY-BROWN, D., ECCLES, J. C., LIDDELL, E. G. T., and SHERRINGTON, C. S.
1938. Reflex activity of the spinal cord. The Clarendon Press, Oxford. 183 pp.

CROISIER, M.
1944. La forme de la contraction musculaire réflexe en fonction de la fréquence des stimulations afférentes. Helv. Physiol. et Pharmacol. Acta *2*: 97–109.

CURE, C., RASMUSSEN, T., and JASPER, H. H.
1948. Activation of seizures and electroencephalographic disturbances in epileptics and in control subjects with "metrazol". Arch. Neurol. Psychiat. *59*: 691–717.

CURTIS, H. J.
1940a. Intercortical connections of corpus callosum as indicated by evoked potentials. J. Neurophysiol. *3*: 407–413.
1940b. Analysis of cortical potentials mediated by corpus callosum. J. Neurophysiol. *3*: 414–422.

DALE, A. S.
1930. Reversed action of chorda tympani on venous outflow from submaxillary gland. J. Physiol. *70*: 449–454.

DALE, H. H.
1937. Transmission of nervous effects by acetylcholine. Harvey Lectures. *32*: 229–245.
1938a. Natural chemical transmitters. Edinburgh Med. J., N.S. *45*: 461–480.
1938b. Acetylcholine as a chemical transmitter of the effects of nerve impulses. The William Henry Welch Lectures. J. M. Sinai Hosp., N.Y. *4*: 401–415 and 416–429.
1938c. Chemical agents transmitting nervous excitations. John Mallet Purser Lecture. Irish J. Med. Sci. (6), *150*: 245–256.

DANDY, W. E.
1927. Experimental investigations on convlusions; their bearing on epilepsy. J. Am. Med. Assoc. *88*: 90–91.
1930. Changes in our conceptions of localization of certain functions in the brain. Proc. Am. Physiol. Soc. 42nd. Annual Meeting. Am. J. Physiol. *93*: 643.

DANDY, W. E., and ELMAN, R.
1925. Experimental epilepsy. Bull. Johns Hopkins Hosp. *36*: 40–49.

DASGUPTA, S. R., MUKHERJEE, K. L., and WERNER, G.
1954. The activity of some central depressant drugs in acute decorticate and diencephalic preparations. Arch. intern. pharmacodynamie. *97*: 149–156.

DAVIS, L. E., and POLLOCK, L. J.
1928. Experimental convulsions: crucial experiment to determine convulsive site. Arch. Neurol. Psychiat. *20*: 756–763.

DAVISON, C., and SCHICK, W.
1935. Spontaneous pain and other subjective sensory disturbances. Research Publs., Assoc. Research Nervous Mental Disease *15*: 457–496.

DELL, P., BONVALLET, M., and HUGELIN, A.
1954. Tonus sympathique, adrénaline et controle réticulaire de la motricité spinale. Electroencephalog. and Clin. Neurophysiol. *6*: 599–618.

DENNY-BROWN, D.
1928. Inhibition as reflex accompaniment of tendon jerk and of other forms of active muscular response. Proc. Roy. Soc. (London), B *103*: 321–336.

DENNY-BROWN, D. E., and LIDDELL, E. G. T.
1927. Stretch reflex as spinal process. J. Physiol. *63*: 144–150.

DESMEDT, J. E.
1949. Les propriétés électrophysiologiques du muscle squelettique au cours de la dégénérescence wallérienne et dans le cas d'une atrophie non wallérienne (résection tendineuse). Arch. intern. physiol. *57*: 98–101.
1959. Physio-pathology of neuromuscular transmission and the trophic influence of motor innervation. Am. J. Phys. Med. *38*: 248–261.

DIAMOND, J. and MILEDI, R.
1959. Sensitivity of fetal and newborn rat muscle to acetylcholine. J. Physiol. *149*: 50P.

DILLE, J. M., and HAZELTON, L. W.
1939. The depressant action of picrotoxin and metrazol. J. Pharmacol. Exp. Therap. *67*: 276–289.

DIRNHUBER, P., and EVANS, C. L.
1954. The effects of anticholinesterases on humoral transmission in the submaxillary gland. Brit. J. Pharm. *9*: 441–458.

DMITRIEV, V. D.
1953. Stability of restored reflex functions. Fiziol. Zhur. S.S.S.R., *39*: 293. Abstracted in J. Mental Sci. *100*: 616, 1954.

DONTAS, A. S., and GESELL, R.
  1955. Intercostal nerve electrical activity. Arch. intern. physiol. et biochem. *63*: 305–317.

DRAKE, C. G.
  1947. The effect of convulsant agents on partially denervated neurones
  of the central nervous system. M.Sc. Thesis, University of Western
  Ontario. London, Canada. 63 pp.

DRAKE, C. G., and MCKENZIE, K. G.
  1953. Mesencephalic tractotomy for pain. J. Neurosurg. *10*: 457–462.

DRAKE, C. G., SEGUIN, J. J., and STAVRAKY, G. W.
  1956. The effect of convulsant agents on partially isolated regions of the
  central nervous system. Can. J. Biochem. and Physiol. *34*: 689–712.

DRAKE, C. G., and STAVRAKY, G. W.
  1948a. Effect of convulsant agents on partially isolated neurones of the
  central nervous system. Federation Proc. *7*: 29.
  1948b. An extension of the "law of denervation" to afferent neurones.
  J. Neurophysiol. *11*: 229–238.

DUSSER de BARENNE, J. G.
  1910a. Die Strychninwirkung auf das Zentralnervensystem (a) Die
  Wirkung des Strychnins auf die Preflextätigkeit der Intervertebral-
  ganglia. Folia neurobiol. Lpz. *4*: 467–474.
  1910b. Die Strychninwirkung auf das Zentralnervensystem (b) Zur
  Wirkung des Strychnins bei localer Application auf das Rücken-
  mark. Folia neurobiol. Lpz. *5*: 42–58.
  1910c. Die Strychninwirkung auf das Zentralnervensystem (c) Die
  segmentare Strychninvergiftung der dorsalen Rückenmarksmecha-
  nismen: ein Beitrag zur Dermatometrie der hinteren Extremität
  des Hundes. Folia neurobiol. Lpz. *5*: 342–359.

DUSSER DE BARENNE, J. G., GAROL, H. W., and MCCULLOCH, W. S.
  1942. Physiological neuronography of the cortico-striatal connections.
  Research Publs., Assoc. Research Nervous Mental Disease *21*:
  246–266.

DUSSER DE BARENNE, J. G., and MCCULLOCH, W. S.
  1939. Physiological delimitation of neurones in the central nervous
  system. Am. J. Physiol. *127*: 620–628.

ECCLES, J. C.
  1941. Changes in muscle produced by nerve degeneration. Med. J.
  Australia *1*: 573–575.
  1953. The neurophysiological basis of mind. The Waynflete Lectures,
  Oxford, 1952. Oxford, The Clarendon Press. 314 pp.
  1955. The central action of antidromic impulses in motor nerve fibers.
  Pflüger's Arch. ges. Physiol. *260*: 385–415.

ECCLES, J. C., FATT, P., and KOKETSU, K.
1954. Cholinergic and inhibitory synapses in a pathway from motor axon collaterals to motoneurones. J. Physiol. *126*: 524–562.

ECCLES, J. C., and McINTYRE, A. K.
1951. Plasticity of mammalian monosynaptic reflexes. Nature *167*: 466–468.
1953. The effects of disuse and of activity on mammalian spinal reflexes. J. Physiol. *121*: 492–516.

ECHLIN, F. A.
1959. The supersensitivity of chronically "isolated" cerebral cortex as a mechanism in focal epilepsy. Electroencephalog. and Clin. Neurophysiol. *11*: 697–722.

ECHLIN, F. A., ARNETT, V., and ZOLL, J.
1952. Paroxysmal high voltage discharges from isolated and partially isolated human and animal cerebral cortex. Electroencephalog. and Clin. Neurophysiol. *4*: 147–164.

ECHLIN, F. A., and McDONALD, J.
1954. The supersensitivity of chronically isolated and partially isolated cerebral cortex as a mechanism in focal cortical epilepsy. Trans. Am. Neurol. Assoc. *79*: 75–79.

ECKENHOFF, J. E., HAFKENSCHIEL, J. H., and LANDMESSER, C. M.
1947. Coronary circulation in dog. Am. J. Physiol. *148*: 582–596.

EDDS, M. F.
1950. Collateral regeneration of residual motor axons in partially denervated muscles. J. Exp. Zool. *113*: 517–551.

ELLIOTT, K. A. C., and FLOREY, E.
1956. Studies on factor I. Proc. Can. Physiol. Soc. *20*: 20–21. Rev. can. biol. *15*: 248–249.

ELSBERG, C. A., and STOOKEY, B. P.
1923. Studies on epilepsy: convulsions experimentally produced in animals compared with convulsive states in man. Arch. Neurol. Psychiat. *9*: 613–626.

EMMELIN, N.
1952a. "Paralytic secretion" of saliva an example of supersensitivity after denervation. Physiol. Revs. *32*: 21–46.
1952b. On the mechanism of paralytic secretion. Acta Physiol. Scand. *26*: 232–242.
1960. Supersensitivity of submaxillary gland following exclusion of the postganglionic parasympathetic neurone. Brit. J. Pharmacol. *15*: 356–360.

154 *References*

EMMELIN, N., JACOBSON, D., and MUREN, A.
   1951. Effects of prolonged administration of atropine and pilocarpine on the submaxillary gland of the cat. Acta Physiol. Scand. *24*: 128–143.

EMMELIN, N., and MUREN, A.
   1950a. Acetylcholine release at parasympathetic synapses. Acta Physiol. Scand. *20*: 13–32.
   1950b. Supersensitivity of denervated organs to chemical stimuli. Nature *166*: 610.
   1951a. Paralytic secretion of saliva. Acta Physiol. Scand. *21*: 362–379.
   1951b. Sensitization of the submaxillary gland to chemical stimuli. Acta Physiol. Scand. *24*: 103–127.
   1952. The sensitivity of submaxillary glands to chemical agents studied in cats under various conditions over long periods. Acta Physiol. Scand. *26*: 221–231.

EMMELIN, N., and STROMBLAD, R.
   1951. Adrenaline and noradrenaline content of the suprarenals of cats in chloralose and morphine-ether anesthesia. Acta Physiol. Scand. *24*: 261–266.

ERICKSON, T. C.
   1940. Spread of epileptic discharge; experimental study of after-discharge induced by electrical stimulation of cerebral cortex. Arch. Neurol. Psychiat. *43*: 429–452.

ERLANGER, J., and GASSER, H. S.
   1930. Action potential in fibers of slow conduction in spinal roots and somatic nerves. Am. J. Physiol. *92*: 43–82.
   1937. Electrical signs of nervous activity. Johnson Foundation Lectures. University of Pennsylvania Press, Philadelphia. 221 pp.

EULER, U. S. V., and GADDUM, J. H.
   1931. Pseudomotor contractures after degeneration of the facial nerve. J. Physiol. *73*: 54–66.

EULER, U. S. V., and PURKHOLD, A.
   1951. Histamine in organs and its relation to the sympathetic nerve supply. Acta Physiol. Scand. *24*: 218–224.

EYCK, M. VAN
   1956. Étude analytique du phénomène de compensation vestibulaire après labyrinthectomie unilatérale. Acta Oto-Laryngol. *46*: 279–284.

FATT, P., and KATZ, B.
   1951. An analysis of the end-plate potential recorded with intra-cellular electrode. J. Physiol. *115*: 320–370.
   1952. Spontaneous subthreshold activity at motor nerve endings. J. Physiol. *117*: 109–128.

FELDBERG, W.
1933. Der nachweis eines acetylcholinähnlichen Stoffes im Zungen-venenblut des Hundes bei Reizung des Nervus lingualis. Pflüger's Arch. ges. Physiol. *232*: 88–104.
1945. Present views on the mode of action of acetylcholine in the central nervous system. Physiol. Revs. *25*: 596–642.

FELDBERG, W., HARRIS, G. W., and LIN, R. C. Y.
1951. Observations on the presence of cholinergic and non-cholinergic neurones in central nervous system. J. Physiol. *112*: 400–404.

FELDBERG, W., and SHERWOOD, S. L.
1954. Injections of drugs into lateral ventrical of cat. J. Physiol. *123*: 148–167.

FELDBERG, W., and VARTIAINEN, A.
1934. Further observations on physiology and pharmacology of sympathetic ganglion. J. Physiol. *83*: 103–128.

FENDER, F. A.
1937. Epileptiform convulsions from "remote" excitation. Arch. Neurol. Psychiat. *38*: 259–267.

FENG, T. P., and TING, Y. C.
1938. Studies on the neuromuscular junction. XI. A note on the local concentration of cholinesterase at motor nerve endings. Chinese J. Physiol. *13*: 141–144.

FERRIER, D.
1873. Experimental researches in cerebral physiology and pathology. West Riding Lunatic Asylum Medical Reports. *3*: 30–96.
1876. The functions of the brain. Smith, Elder and Co., London, 323 pp.

FISCHER, M. H.
1932. Aktionsströme der Hirnrinde. XIV Cong. Intern. di Fisiol., Roma. Sunti delle communicazioni scientifiche. Bologna, Cappelli, pp. 80–81.

FISHER, S. M., and STAVRAKY, G. W.
1944. The effects of acetyl-beta-methylcholine in human subjects with localized lesions of the central nervous system. Am. J. Med. Sci. *208*: 371–380.

FLEISCH, A.
1931. Die Wirkung von Histamine, Acetylcholin und Adrenalin auf die Venen. Pflüger's Arch. ges. Physiol. *228*: 351–372.

FLEMING, A. J., and MACINTOSH, F. C.
1935. The effect of sympathetic stimulation and of autonomic drugs on the paralytic submaxillary gland of the cat. Quart. J. Exp. Physiol. *25*: 207–212.

FLOREY, E.

1953. Ueber einen nervösen Hemmungsfaktor in Gehirn und Rücken-mark. Naturwissenschaften *40*: 295–296.

1954a. Inhibition and excitatory factor of mammalian central nervous system, and their action on single sensory neuron. Arch. intern. physiol. *62*: 33–53.

1954b. Ueber die Wirkung von Acetylcholine, Adrenalin, Nor-adrena-lin, Faktor E und anderen Substanzen auf den isolierten Enddarm des Flusskrebsen Cambarus Clarkii Girard. Z. vergleich. Physiol. *36*: 1–8.

1956. The action of factor I on certain invertebrate organs. Can. J. Biochem. Physiol. *34*: 669–681.

FLOREY, E., and McLENNAN, H.

1955a. Die Wirkung des Hemmungsfaktors auf Gehirn und Rücken-mark auf periphere und zentral synaptische Uebertragung. Natur-wissenschaften *42*: 51–52.

1955b. The release of an inhibitory substance from mammalian brain, and its effect on peripheral synaptic transmission. J. Physiol. *129*: 384–392.

1955c. Effects of an inhibitory factor (Factor I) from brain and central synaptic transmission. J. Physiol. *130*: 446–455.

FOERSTER, O.

1918. Die operative Behandlung der spastischen Lähmungen. (Hemi-plegie, Monoplegie, Paraplegie) bei Kopf-und Rückenmarkschüs-sen. Deut. Z. Nervenheilk. *58*: 151–215.

FOGGIE, P.

1937. Action of adrenaline, acetylcholine, and histamine on lungs of rat. Quart. J. Exp. Physiol. *26*: 225–233.

FOLKOW, B.

1955. Nervous control of the blood vessels. Physiol. Revs. *35*: 629–663.

FOLKOW, B., FROST, J., and UNVAS, B.

1948. Action of adrenaline, noradrenaline and some other sympatho-mimetic drugs on muscular, cutaneous and splanchnic vessels of cat. Acta Physiol. Scand. *15*: 412–420.

1949. Action of acetylcholine, adrenaline and noradrenaline on coro-nary blood flow of dog. Acta Physiol. Scand. *17*: 201–205.

FORSTER, F. M., and MADOW, L.

1950. Metrazol activation of acetylcholine-treated cerebral cortex. Am. J. Physiol. *161*: 426–434.

FRANK, E., NOTHMANN, N. M., and HIRSCHMAN-KAUFMANN, H.

1922. Ueber die 'tonische' Kontraktion des quergestreiften Säugetiermus-kels nach Ausschaltung des motorischen Nerven. Pflüger's Arch. ges. Physiol. *197*: 270–287.

FRANK, G. B.
1959. Increased sensitivity of an end-plate region to its chemical mediator following denervation of another end-plate region of same cell. Can. J. Biochem. Physiol. *37*: 1239–1246.

FRENCH, J. D., VERZEANO, M., and MAGOUN, H. W.
1953. A neural basis of the anaesthetic state. A.M.A. Arch. Neurol. Psychiat. *69*: 519–529.

FRIEDLANDER, H. D., PERLMAN, I., and CHAIKOFF, I. L.
1941. The effects of denervation on phospholipid activity of skeletal muscle as measured with radioactive phosphorus. Am. J. Physiol. *132*: 24–31.

FRITSCH, G., and HITZIG, E.
1870. Ueber die elektrische Erregbarkeit des Grosshirns. Arch. Anat. u. Physiol. *37*: 300–332.

FRÖHLICH, F. W.
1909. Beiträge zur Analyse der Reflexfunktion des Rückenmarks mit besonderer Berücksichtigung von Tonus, Bahnung und Hemmung. Z. allgem. Physiol. *9*: 55–111.

FRÖHLICH, A., and LÖEWI, O.
1906. Ueber vasokonstriktorische Fasern in der Chorda tympani. Zentr. Physiol. *20*: 229–232.
1908. Physiologie des zentralen und sympathischen Nervensystems. Zentr. Physiol. *22*: 491–492.

FROMMEL, E., and RADOUCO-THOMAS, C.
1953. Le point d'impact de la diphenylhydantoine du phénobarbital et des acylurées, peut-il nous servir à la compréhension du mécanisme physiolpathologique de l'épilepsie? Helv. Physiol. et Pharmacol. Acta *11*: 231–238.

FULTON, J. F.
1926. Muscular contraction and the reflex control of movement. Williams and Wilkins Co., Baltimore. 644 pp.
1949. Physiology of the nervous system. 3rd ed. Oxford University Press, New York. 667 pp.

FULTON, J. F., and KELLER, A. D.
1932. The sign of Babinski. A study of the evolution of cortical dominance in primates. C. C Thomas, Springfield, Ill. 165 pp.

FULTON, J. F., and LIDDELL, E. G. T.
1925. Electrical responses of extensor muscles during postural (myotatic) contraction. Proc. Roy. Soc. London, B *98*: 577–589.

FULTON, J. F., LIDDELL, E. G. T., and RIOCH, D. M.
1930. Influence of experimental lesions of spinal cord upon knee-jerk; acute lesions. Brain *53*: 311–326.

FULTON, J. F., and McCOUCH, G. P.
  1937. Relation of motor area of primates to hyporeflexia (spinal shock) of spinal transection. J. Nervous Mental Disease *86*: 125–146.

FULTON, J. F., and PI-SUÑER, J.
  1928. Note concerning probable function of various afferent end-organs in skeletal muscle. Am. J. Physiol. *83*: 554–562.

GABBE, E.
  1926. Ueber die Wirkung der sympathischen Innervation auf die Zirkulation und den Stoffaustausch in den Muskeln. Z. ges. exp. Med. *51*: 728–751.

GADDUM, J. H., and HOLTZ, P.
  1933. Localization of action of drugs on pulmonary vessels of dogs and cats. J. Physiol. *77*: 139–158.

GAROL, H. W.
  1942. Cortical origin and distribution of corpus callosum and anterior commissure in cat. J. Neuropathol. Exp. Neurol. *1*: 422–429.

GARREY, W. E.
  1924. Auricular fibrillation. Physiol. Revs. *4*: 215–250.

GASSER, H. S.
  1934. Conduction in nerves in relation to fiber types. Research Publs., Assoc. Research Nervous Mental Disease *15*: 35–59.
  1937. Control of excitation in nervous system. Harvey Lectures. *32*: 169–193.

GELFAN, S., and TARLOV, I. M.
  1959. Interneurones and rigidity of spinal origin. J. Physiol. *146*: 594–617.

GELLHORN, E.
  1953. Physiological foundations of neurology and psychiatry. University of Minnesota Press, Minneapolis. 556 pp.

GELLHORN, E., DARROW, C. W., and YESINICK, L.
  1939. Effect of epinephrine on convulsions. Arch. Neurol. Psychiat. *42*: 826–836.

GELLHORN, E., and MINATOYA, H.
  1943. Effect of insulin hypoglycemia on conditioned reflexes. J. Neurophysiol. *6*: 161–171.

GERARD, R. W.
  1941. The interaction of neurones. Ohio J. Sci. *41*: 160–172.
  1951. Physiology of pain; abnormal neurone states in causalgia and related phenomena. Anesthesiology *12*: 1–13.

GERGENS, E.
  1876. Ueber gekreuzte Reflexe. Pflüger's Arch. ges. Physiol. *14*: 340–344.

GERNANDT, B. E., and TERZUOLO, C. A.
  1955a. Effect of vestibular stimulation upon activity of spinal cord. Federation Proc. *14*: 56.
  1955b. Effect of vestibular stimulation on strychnine-induced activity of the spinal cord. Am. J. Physiol. *183*: 1–8.

GESELL, R.
  1919. Studies on the submaxillary gland. III. Factors controlling volume-flow of blood. Am. J. Physiol. *47*: 438–467.
  1920. Studies on submaxillary gland. VI. On the dependence of tissue activity upon volume-flow of blood and on the mechanism controlling this volume-flow of blood. Am. J. Physiol. *54*: 166–184.
  1940. A neurophysiological interpretation of the respiratory act. Ergeb. Physiol. *43*: 477–639.

GESELL, R., and HANSEN, E. T.
  1943. Eserine, acetylcholine, atropine and nervous integration. Am. J. Physiol. *139*: 371–385.

GIACHETTI, A., PERUZZI, P., and SPADOLINI, I.
  1953. Acetylcholine, diphasic action on cholinergic mechanism of cardiac regulation. XIX Intern. Physiol. Congr., Montreal, p. 391.

GIBBS, F. A., DAVIS, H., and LENNOX, W. G.
  1935. Electroencephalogram in epilepsy and in conditions of impaired consciousness. Arch. Neurol. Psychiat. *34*: 1133–1148.

GIRNDT, O.
  1926. Physiologische Beobachtungen an Thalamuskatzen. Pflüger's Arch. ges. Physiol. *213*: 427–486.

GLEES, P., and CLARK, W. E. L.
  1941. The termination of optic fibres in the geniculate body of the monkey. J. Anat. *75*: 295–308.

GOETZ, R. H.
  1939. Control of blood-flow through intestine as studied by effect of adrenaline. Quart. J. Exp. Physiol. *29*: 321–332.

GOLDSTEIN, H. H., and WEINBERG, J.
  1940. Metrazol as diagnostic aid in epilepsy. Am. J. Psychiat. *96*: 1455–1458.

GOODALL, M.
  1951. Studies of adrenaline and noradrenaline in mammalian heart and suprarenals. Acta Physiol. Scand. *24*: Suppl. 85, 1–51.

GOODMAN, L. S., GREWAL, M. S., BROWN, W. C., and SWINYARD, E. A.
1953. Comparison of maximal seizures evoked by pentylenetetrazol (metrazol) and electroshock in mice and their modification by anticonvulsants. J. Pharmacol. Exp. Therap. *108*: 168–176.

GOODMAN, L. S., SWINYARD, E. A., and TOMAN, J. E.
1946. Effects of 1 (+) glutamic acid and other agents on experimental seizures. Arch. Neurol. Psychiat. *56*: 20–29.

GOODWIN, J. E., KERR, W. K., and LAWSON, F. L.
1940. Bioelectric responses in metrazol and insulin shock. Am. J. Psychiat. *96*: 1389–1405.

GOZZANO, M.
1936. Bioelectrische Erscheinungen bei der Reflexepilepsi. J. Psychol. u. Neurol. *47*: 24–39.

GRAFSTEIN, B.
1957. Response of cerebral cortex to stimulation of a point in the opposite hemisphere. J. Physiol. *137*: 40 P.
1959. Organization of callosal connections in suprasylvian gyrus of cat. J. Neurophysiol. *22*: 504–515.

GRAHAM, A. R.
1954. Reflex patterns in the hind limb of the cat: effects of sensitization by chronic denervation, of labyrinthine stimulation and of muscle loading upon them. Ph.D. Thesis. University of Western Ontario, London, Canada. 187 pp.

GRAHAM, A. R., and STAVRAKY, G. W.
1953a. Reversal of the effects of chorda tympani stimulation, and of acetylcholine and adrenaline, as seen in the submaxillary salivary gland of the cat. Rev. can. biol. *11*: 446–470.
1953b. Effect of variation in intensity of afferent stimulation of the reflex response of antagonists about the ankle joint in decerebrate and spinal cat. Federation Proc. *12*: 55.
1954a. Studies of reciprocal innervation in spinal and decerebrate cats sensitized by preceding semisection of spinal cord or by semide-cerebration. Federation Proc. *13*: 59–60.
1954b. The effects of chronic semisection of the spinal cord on reflexes of antagonists in the shank of the cat. Rev. can. biol. *13*: 466.
1954c. The response of the chronically denervated submaxillary gland to acetylcholine and to adrenaline. Rev. can. biol. *13*: 120–143.

GRANIT, R.
1950. Reflex self-regulation of the muscle contraction and autogenetic inhibition. J. Neurophysiol. *13*: 351–372.
1955. Receptors and sensory perception: a discussion of aims, means, and results of electrophysiological research into the process of reception. Silliman Memorial Lectures. Yale University Press, New Haven. 369 pp.

GRANIT, R., and KAADA, B. R.
1952. Influence of stimulation of central nervous structures on muscle spindles in cat. Acta. Physiol. Scand. *27*: 130–160.

GRINKER, R. R., and BUCY, P. C.
1949. Neurology. 4th ed. C. C Thomas, Springfield, Ill. 1138 pp.

HAGBARTH, K. E.
1952. Excitatory and inhibitory skin areas for flexor and extensor motoneurones. Acta Physiol. Scand. *26*: Suppl. 94, 1–58.
1953. Specific skin areas for excitation and inhibition of hindlimb reflexes. Ch. 18, pp. 261–271. *In* MALCOLM, J. L., and GRAY, J. A. B., eds., The Spinal Cord. J. & A. Churchill, London. 300 pp.

HAHN, F.
1941a. Zur Metodik des Nachweises der zentralanaleptischen Blut-drucksteigerung. Arch. exp. Pathol. Pharmakol. *198*: 473–490.
1941b. Vergleichende Untersuchungen über die Krampf—und Blut-druckwirkung verschiedener Analeptica. Arch. exp. Pathol. Phar-makol. *198*: 491–508.
1941c. Vergleichende Untersuchungen über die Krampf—und Blut-druckwirkung einiger Analeptica an dekapitierten Katzen. Arch. exp. Pathol. Pharmakol. *198*: 509–526.

HALL, G. E.
1938. Physiological studies in experimental insulin and metrazol shock: composite preliminary study by members of Department of Medical Research Banting Institute, University of Toronto. Am. J. Psychiat. *18*: 553–566.

HALL, M.
1841. On the diseases and derangements of the nervous system. Bail-lière, London. *Cited by* RIESE, W., Principles of neurology in the light of history and their present use, p. 97. Nervous and Mental Disease Monographs, New York, 1950. 177 pp.

HALPERN, L., and SHARPLESS, S.
1959. Denervation supersensitivity in undercut cortex studied by means of chronically implanted electrodes. The Pharmacologist *1*: 66.

HAMPEL, C. W.
1935. The effect of denervation on the sensitivity to adrenine of the smooth muscle in the nictitating membrane of the cat. Am. J. Physiol. *111*: 611–621.

HANSEN, E. T., WORZNIAK, J. J., and GESELL, R.
1942. Physiological effects of artificially administered acetylcholine and eserine. Federation Proc. *1*: 36.

HARPUDER, K., BYER, J., and STEIN, I. D.
1947. The effect of intra-arterial injection of adrenalin upon blood flow of the human forearm. Am. J. Physiol. *150*: 181–189.

HARREVELD, A. VAN
  1940. On spinal shock. Am. J. Physiol. *129*: 515–523.
  1943. Tone and tendon reflexes after asphyxiation of the spinal cord. Am. J. Physiol. *139*: 615–625.
  1944. Reflexes in the anterior tibial muscle after cord asphyxiation. Am. J. Physiol. *142*: 428–434.
  1947. On mechanism and localization of symptoms of electro-shock and electro-narcosis. J. Neuropathol. Exp. Neurol. *6*: 177–184.

HARREVELD, A. VAN, and MARMONT, G.
  1939. The course of recovery of the spinal cord from asphyxia. J. Neurophysiol. *2*: 101–111.

HARREVELD, A. VAN, and STAMM, J. J.
  1954. Consequences of cortical convulsive activity in the rabbit. J. Neurophysiol. *17*: 505–520.

HARRIS, E. J., and NICHOLLS, J. G.
  1956. The effect of denervation on the rate of entry of potassium into frog muscle. J. Physiol. *131*: 473–476.

HATTORI, H.
  1936. Ueber die Einwirkung verschiedener Gefässmittel auf die Schild-drüsengefässe des Rindes. Folia Endocrinol. Japon. *11*: 62–63.

HEAD, H.
  1921. Release of function in the nervous system. Croonian Lecture. Proc. Roy. Soc. London *B 92*: 184–208.

HEAD, H., and HOLMES, G.
  1911. Sensory disturbances from cerebral lesions. Brain *34*: 102–254.

HEAD, H., RIVERS, W. H. R., and SHERREN, J.
  1905. The afferent nervous system from a new aspect. Brain *28*: 99–115.

HEBB, C. O.
  1957. Biochemical evidence for the neural function of acetylcholine. Physiol. Revs. *37*: 196–220.

HENDERSON, V. E., and ROEPKE, M. H.
  1933a. On mechanism of salivary secretion. J. Pharmacol. Exp. Therap. *47*: 193–207.
  1933b. Ueber den lokalen hormonalen Mechanismus der Parasympathi-kusreizung. Arch. exp. Pathol. Pharmakol. *172*: 314–324.

HENRY, C. E., and SCOVILLE, W. B.
  1952. Suppression-burst activity from isolated cerebral cortex in man. Electroencephalog. and Clin. Neurophysiol. *4*: 1–22.

HERRICK, C. J.
  1956. The evolution of human nature. University of Texas Press, Austin. 506 pp.

References   163

HESS, W. R.
1947. Die Relativität biologischer Reactionen. Festschrift A. Stoll. B. Schwabe and Co., Basel. 655 pp., p. 3–9.

HINSEY, J. C., RANSON, S. W., and DIXON, H. H.
1930. Responses elicited by stimulation of mesencephalic tegmentum in cat. Arch. Neurol. Psychiat. *24*: 966–977.

HINSEY, J. C., RANSON, S. W., and DOLES, E. A.
1930. Reversal in crossed-extension reflex in decerebrate, decapitate and spinal cats. Am. J. Physiol. *95*: 573–583.

HINSEY, J. C., RANSON, S. W., and MCNATTIN, R. F.
1930. Role of hypothalamus and mesencephalon in locomotion. Arch. Neurol. Psychiat. *23*: 1–43.

HIROSE, Y.
1932. Ueber die Wirkung grosser Azetylcholindosen auf die Darm-, Nieren-, Lungen- und Extremitätengefässe. Arch. exp. Pathol. Pharmakol. *165*: 401–406.

HOEFER, R. F. A., and POOL, J. L.
1943. Conduction of cortical impulses and motor management of convulsive seizures. Arch. Neurol. Psychiat. *50*: 381–400.

HOFFMAN, H.
1950. Local re-innervation in partially denervated muscle: a histophysiologic study. Australian J. Exp. Biol. Med. Sci. *28*: 383–397.

HORSLEY, VICTOR
1907. Dr. Hughlings Jackson's views of the functions of the cerebellum. The Hughlings Jackson Lecture for 1906. Brit. Med. J. *1*: 803–808.

HUGHES, R. A.
1950. A study of the inhibitory processes in neurones sensitized by partial isolation. M.Sc. Thesis. University of Western Ontario, London, Canada. 140 pp.

HUGHES, R. A., STAVRAKY, G. W., and TEASDALL, R. D.
1950. A study of the mechanism of reversal of spinal and bulbar reflexes. Trans. Roy. Soc. Can., 3rd series, Sect. V, *44*: 39–60.

HUMOLLER, F. L., HATCH, D., and MCINTYRE, A. R.
1951. Effect of neurotomy on hexokinase and phosphorylase activity of rat muscle. Am. J. Physiol. *167*: 656–664.
1952. Cytochrome oxidase activity in muscle following neurotomy. Am. J. Physiol. *170*: 371–374.

HUNT, C. C.
1951. The reflex activity of mammalian small-nerve fibres. J. Physiol. *115*: 456–469.

1952a. The effect of stretch receptors from muscle on the discharge of motoneurones. J. Physiol. *117*: 359–379.

1952b. Peripheral origins of nervous activity. Muscle stretch receptors; peripheral mechanisms and reflex function. Cold Spring Harbor Symposia. Quant. Biol. *17*: 113–123.

HUNT, C. C., and KUFFLER, S. W.

1951a. Further study of efferent small-nerve fibres to mammalian muscle spindles. Multiple spindle innervation and activity during contraction. J. Physiol. *113*: 283–297.

1951b. Stretch receptor discharges during muscle contraction. J. Physiol. *113*: 298–315.

HUNT, J. R.

1917. Progressive atrophy of the globus pallidus (primary atrophy of the pallidal system). Contribution to the functions of the corpus striatum. Brain *40*: 58–148.

1918. Primary atrophy of the pallidal system of the corpus striatum. Arch. Internal Med. *22*: 647–691.

1923. The dual nature of the efferent nervous system. Arch. Neurol. Psychiat. *10*: 37–82.

1931. A theory of the mechanism underlying inhibition in the central nervous system and its relation to convulsive manifestations. Research Publs., Assoc. Research Nervous Mental Disease *7*: 45–64.

HUNT, R.

1918. Vasodilator reactions. Am. J. Physiol. *45*: 197–231.

IRONSIDE, R., and GUTTMACHER, M.

1929. The corpus callosum and its tumors. Brain *52*: 442–483.

IVY, A. C., GOETZL, F. R., HARRIS, S. C., and BURRILL, D.Y.

1944. Analgesic effect of intracarotid and intravenous injection of epinephrine in dogs and of subcutaneous injections in man. Quart. Bull. Northwestern Univ. Med. School *18*: 298–306.

JACKSON, J. H.

1870. A study of convulsions. Trans. St. Andrews Med. Grad. Assoc. *3*: 1–45. *Reprinted in* Selected writings of John Hughlings Jackson, ed. J. TAYLOR. Hodder and Stoughton, London. 1931. 500 pp., v. I, pp. 8–36.

1884. Evolution and dissolution of the nervous system. The Croonian Lectures. Brit. Med. J. *1*: 591–593, 660–663, 703–707.

JENKNER, F. L., and WARD, A. A., JR.

1953. Bulbar reticular formation and tremor. A. M. A. Arch. Neurol. Psychiat. *70*: 489–502.

KAADA, B. R.

1953. (Postural) suprasegmental mechanisms. XIX Intern. Physiol. Congr., Montreal, pp. 81–87.

KABAT, H., and GRENELL, R. G.
  1944. Unpublished observations. *Quoted from* KABAT, H., and KNAPP, M. E., The mechanism of muscular spasm in poliomyelitis. J. Pediat. *24*: 123–137 (1944).

KALINOWSKY, L. B.
  1947. Epilepsy and convulsive therapy. Research Publs., Assoc. Research Nervous Mental Disease *26*: 175–183.

KALINOWSKY, L. B., and KENNEDY, F.
  1941. Theories of epilepsy in the light of electric shock therapy. Trans. Am. Neurol. Assoc. *67*: 194–197.

KARPLUS, I. P.
  1914. Experimenteller Beitrag zur Kenntnis der Gehirnvorgänge beim epileptischen Anfall. Wien. klin. Wochschr. *27*: 645–651.

KATZ, L., LINDNER, E., WEINSTEIN, W., ABRAMSON, D. I., and JOCHIM, K.
  1938. Effects of various drugs on coronary circulation of denervated isolated heart of dog and cat, observations on ephinephrine, acetylcholine, acetyl-B-methylcholine, nitroglycerine, sodium nitrate, pitressin and histamine. Arch. intern. pharmacodynamie *59*: 399–415.

KATZ, L., JOCHIM, K., and BOHNING, A.
  1938. Effect of extravascular support of ventricles on flow in coronary vessels. Am. J. Physiol. *122*: 236–251.

KATZ, L., and JOCHIM, K.
  1939. Observations on innervation of coronary vessels of dog. Am. J. Physiol. *126*: 395–401.

KAUFMAN, I. C., MARSHALL, C., and WALKER, A. E.
  1947. Metrazol-activated electroencephalography. Research Publs., Assoc. Research Nervous Mental Disease *26*: 476–486.

KEITH, H. M.
  1933. Factors influencing experimentally produced convulsions. Arch. Neurol. Psychiat. *29*: 148–154.

KEITH, H. M., and STAVRAKY, G. W.
  1935. Experimental convulsions induced by administration of thujone. Arch. Neurol. Psychiat. *34*: 1022–1040.

KENDALL, D.
  1939. Some observations on central pain. Brain *62*: 253–273.

KENNARD, M. A.
  1953. Sensitization of the spinal cord of the cat to pain-inducing stimuli. J. Neurosurg. *10*: 169–177.

KENNARD, M. A., and WATTS, J. W.
  1934. Effect of section of corpus callosum on motor performance of monkeys. J. Nervous Mental Disease *79*: 159–169.

KESSLER, M., and GELLHORN, E.
1943. Effect of electrically and chemically induced convulsions on conditioned reflexes. Am. J. Psychiat. *99*: 687–691.

KILLAM, E. K., and KILLAM, K. F.
1957. Brain mechanisms and drug action. FIELDS, W. S., ed. C. C Thomas, Springfield, Ill., pp. 71–98.

KING, E. E.
1956. Differential action of anesthetics and interneuron depressants upon EEG arousal and recruitment responses. J. Pharmacol. Exp. Therap. *116*: 404–417.

KING, E. E., NAQUET, R., and MAGOUN, H. W.
1957. Alterations in somatic afferent transmission through the thalamus by central mechanisms and barbiturates. J. Pharmacol. Exp. Therap. *119*: 48–63.

KING, L. S., and MEEHAN, M. C.
1936. Primary degeneration of corpus callosum (Marchiafava's disease) Arch. Neurol. Psychiat. *36*: 547–568.

KIRILLOVA, A. A.
1943. The effect of acetylcholine on the isolated spinal cord of the frog. Byulletin eksperimentalnoi biologii i meditsiny *16*: 53–56. *Abstracted* Am. Rev. Soviet Med. *4*: 269 (1946–47).

KLEIJN, A. de
1923. Experimental physiology of labyrinth. J. Laryngol. and Otol. *38*: 646–663.

KNOEFEL, P. K., and MURRELL, F. C.
1935. Action of some stimulants on spinal reflexes. Arch. intern. pharmacodynamie *52*: 48–53.

KNOWLTON, G. C., and HINES, H. M.
1937. Acetylcholine contracture of denervated muscle. Am. J. Physiol. *120*: 757–760.

KOHN, R., LEVITSKY, P., STRAUSS, A. A., STRAUSS, S., and NECHELES, H.
1936. The vasoconstrictor effect of acetylcholine on isolated splanchnic blood-vessels of dog and man. Arch. intern. pharmacodynamie *53*: 421–425.

KONRADI, G. P.
1958. Activity of the central nervous system. Chapter 54. Coordination of reflex activity and central inhibition. *In* BYKOV, K. M. (ed.), VLADIMIROV, G. Y., DELOV, V. Y., KONRADI, G. P., and SLONIM, A. D. Textbook of Physiology. Translated by BELSKY, S. and MYSHNE, D. Foreign Languages Publishing House, Moscow. 763 pp.

KOPELOFF, N., KENNARD, M. A., PACELLA, B. L., KOPELOFF, L. M., and CHUSID, J. G.
1950. Section of corpus callosum in experimental epilepsy in the monkey. Arch. Neurol. Psychiat. *63*: 719–727.

KRISTIANSEN, K., and COURTOIS, G.
1949. Rhythmic electrical activity from isolated cerebral cortex. Electroencephalog. and Clin. Neurophysiol. *1*: 265–272.

KUFFLER, S. W.
1943. Specific excitability of the endplate region in normal and denervated muscle. J. Neurophysiol. *6*: 99–110.
1953. (Postural) segmental mechanisms. XIX Intern. Physiol. Congr., Montreal, pp. 76–81.

KUFFLER, S. W., and HUNT, C. C.
1952. The mammalian small-nerve fibers: a system for efferent nervous regulation of muscle spindle discharge. Research Publs., Assoc. Research Nervous Mental Disease *30*: 24–27.

KUFFLER, S. W., HUNT, C. C., and QUILLIAM, J. P.
1951. Function of medullated small-nerve fibers in mammalian ventral roots: efferent muscle spindle innervation. J. Neurophysiol. *14*: 29–54.

LANDAU, W. M., and BISHOP, G. H.
1953. Pain from dermal, periosteal, and fascial endings and from inflammation; electrophysiological study employing differential nerve blocks. A.M.A. Arch. Neurol. Psychiat. *69*: 490–504.

LAPORTE, Y., and LLOYD, D. P. C.
1952. Nature and significance of the reflex connections established by large afferent fibers of muscular origin. Am. J. Physiol. *169*: 609–621.

LAPORTA, M., and SAVIANO, M.
1943. Azione dell' acetilcolina sui preparato cardiopulmonare di cane. Arch. fisiol. *43*: 22–39.

LASSEK, A. M.
1946. Pyramidal tract: representation of lateral corticospinal component in spinal cord of cat. J. Neuropathol. Exp. Neurol. *5*: 72–76.

LEIMDORFER, A., and METZNER, W. R. T.
1949. Analgesia and anesthesia induced by epinephrine. Am. J. Physiol. *157*: 116–121.

LEKSELL, L.
1945. The action potential and excitatory effects of the small ventral root fibres to skeletal muscle. Acta Physiol. Scand. *10*: Suppl. 31, 1–88.

LERICHE, R.
    1939. The surgery of pain. Translated and edited by A. YOUNG. Williams and Wilkins Co., Baltimore. 512 pp.

LEWIS, THOMAS
    1925. The mechanism and graphic registration of the heart beat. 3rd ed. Shaw and Sons, London. 329 pp.
    1942. Pain. Macmillan Co., New York. 192 pp.

LEWIS, T., and POCHIN, E. E.
    1937. Double pain response of human skin to single stimulus. Clin. Sci. *3*: 67–76.
    1938. Effects of asphyxia and pressure on sensory nerves of man. Clin. Sci. *3*: 141–155.

LI, C.
    1960. Mechanism of fibrillation potentials in denervated mammalian skeletal muscle. Science *132*: 1889–1890.

LIBET, B., FEINSTEIN, B., and WRIGHT, E. W., JR.
    1955. Tendon afferents in autogenetic inhibition. Federation Proc. *14*: 92.

LIDDELL, D. W., and RETTERSTÖL, N.
    1957. The occurrence of epileptic fits in leucotomized patients receiving chlorpromazine therapy. J. Neurol. Neurosurg. Psychiat. *20*: 105–107.

LIDDELL, E. G. T.
    1934. Spinal shock and some features in isolation-alteration of spinal cord in cats. Brain *57*: 386–400.
    1936. Influence of experimental lesions of spinal cord upon knee-jerk: chronic lesions. With appendix, "Note on 'spinal' and 'decerebrate' type of knee-jerk in cat." Brain *59*: 160–174.

LIDDELL, E. G. T., and SHERRINGTON, C. S.
    1924. Reflexes in response to stretch (myotatic reflexes) Proc. Roy. Soc. London *B 96*: 212–242.
    1925. Further observations on myotatic reflexes. Proc. Roy. Soc. London *B 97*: 267–283.

LIEBERT, E., and WEIL, A.
    1939. Histopathology of brain following metrazol injections. Elgin Papers. *3*: 51–57.

LIEPMANN, H., and MAAS, O.
    1908: Fall von linksseitiger Agraphie und Apraxie bei rechtsseitiger Lähmung. J. Psychol. u. Neurol. *10*: 214–227.

LILJESTRAND, G., and MAGNUS, R.
    1919. Ueber die Wirkung des Novokains auf den normalen und den tetanusstarren Skelettmuskel und über die Entstehung der lokalen Muskelstarre beim Wundstarrkrampf. Pflüger's Arch. ges. Physiol. *176*: 168–208.

LINDSLEY, D. B.
1952. Brain stem influences on spinal motor activity. Research Publs., Assoc. Research Nervous Mental Disease *30*: 174–237.

LINDSLEY, D. B., SCHREINER, L. H., KNOWLES, W. B., and MAGOUN, H. W.
1950. Behavioural and EEG changes following chronic brain stem lesions in the cat. Electroencephalog. and Clin. Neurophysiol. *2*: 483–498.

LISSMAN, H. W.
1950. Proprioceptors. *In* Physiological mechanisms in animal behaviour. Symposia Soc. Exp. Biol., IV. University Press, Cambridge. 482 pp. pp. 34–59.

LIVINGSTON, W. K.
1947. Pain mechanisms: a physiologic interpretation of causalgia and its related states. Macmillan Co., New York. 253 pp.

LLOYD, D. P. C.
1941a. Activity in neurones of the bulbospinal correlation system. J. Neurophysiol. *4*: 115–134.
1941b. The spinal mechanism of the pyramidal system in cats. J. Neurophysiol. *4*: 525–546.
1943a. Reflex action in relation to pattern and peripheral source of afferent stimulation. J. Neurophysiol. *6*: 111–119.
1943b. Neuron patterns controlling transmission of ipsilateral hind limb reflexes in cat. J. Neurophysiol. *6*: 293–315.
1943c. Conduction and synaptic transmission of reflex response to stretch in spinal cats. J. Neurophysiol. *6*: 317–326.
1944. Functional organization of the spinal cord. Physiol. Revs. *24*: 1–17.
1946a. Facilitation and inhibition of spinal motoneurons. J. Neurophysiol. *9*: 421–438.
1946b. Integrative pattern of excitation and inhibition in two neuron reflex arcs. J. Neurophysiol. *9*: 439–444.

LLOYD, D. P. C., and CHANG, H. T.
1948. Afferent fibers in muscle nerves. J. Neurophysiol. *11*: 199–208.

LOEWI, O.
1944. Acetylcholine and transmission of nerve impulses. J. Am. Med. Assoc. *124*: 37–38.
1945a. Chemical transmission of nerve impulse. Am. Scientist *33*: 159–174.
1945b. Aspects of transmission of nervous impulse; mediation in peripheral and central nervous system. Edward Gamaliel Janeway Lecture. J. M. Sinai Hosp., N.Y. *12*: 803–816.
1945c. Aspects of the transmission of the nervous impulse: theoretical and clinical implications. J. M. Sinai Hosp., N.Y. *12*: 851–865.
1949. On the hypersensitivity of denervated structures. Confinia Neurologica *9*: 58–63.

LOMAS, J., BOARDMAN, R. H., and MARKOWE, M.
  1955. Complications of chlorpromazine therapy in 800 mental hospital patients. Lancet *1*: 1144–1147.

LORENTE DE NO, R.
  1939. Transmission of nerve impulses through cranial motor nuclei. J. Neurophysiol. *2*: 402–464.

LUBSEN, N.
  1941a. Experimental studies on cerebral circulation of unanaesthetized rabbit; action of adrenaline. Arch. néerl. physiol. *25*: 306–322.
  1941b. Experimental studies on the cerebral circulation of the unanaesthetized rabbit. Arch. néerl. physiol. *25*: 361–365.

LUCO, J. V., and EYZAGUIRRE, C.
  1952. Inhibitory influences on sensitivity of spinal cord neurones. Acta Physiol. Latinoam. *2*: 33–42.

LUNDBERG, A.
  1952. Adrenaline and transmission in sympathetic ganglion of cat. Acta Physiol. Scand. *26*: 252–263.

LYMAN, C. P.
  1942. Penetration of radioactive potassium in denervated muscle. Am. J. Physiol. *137*: 392–395.

MACINTOSH, F. C.
  1937. Choline-esterase content of normal and denervated submaxillary gland of cat. Proc. Soc. Exp. Biol. Med. *37*: 248–251.

MACKAY, M. E.
  1927. Vascular reaction of pilocarpinized submaxillary gland to histamine. J. Pharmacol. Exp. Therap. *32*: 147–159.

McCOUCH, G. P.
  1924. Relations of pyramidal tract to spinal shock. Am. J. Physiol. *71*: 137–152.

McCOUCH, G. P., AUSTIN, G. M. LIU, C. N., and LIU, C. Y.
  1958. Sprouting as a cause of spasticity. J. Neurophysiol. *21*: 205–216.

McCOUCH, G. P., DEERING, I. D., and LING, T. H.
  1951. Location of receptors for tonic neck reflexes. J. Neurophysiol. *14*: 191–195.

McCOUCH, G. P., DEERING, I. D., and STEWART, W. B.
  1950. Inhibition of knee jerk from tendon spindles of crureus. J. Neurophysiol. *13*: 343–350.

McCOUCH, G. P., HUGHES, J., and STEWART, W. B.
  1943. Monkey (macaca mulatta) after hemisection and subsequent transection of spinal cord. J. Neurophysiol. *6*: 155–159.

MCCULLOCH, W. S., and GAROL, H. W.
1941. Cortical origin and distribution of corpus callosum and anterior commissure in monkey (macaca mulatta). J. Neurophysiol. *4*: 555–563.

MCDOWALL, R. J. S.
1946. Stimulating action of acetylcholine on heart. J. Physiol. *104*: 392–403.

MCINTYRE, A. K.
1953. Synaptic function and learning. XIX Intern. Physiol. Congr., Montreal, pp. 107–114.

MCKENZIE, F. A., SEGUIN, J. J., and STAVRAKY, G. W.
1960. Effects of frontal lobectomy upon electrically induced convulsions and electronarcosis. A.M.A. Arch. Neurol. *2*: 55–61.

MCLENNAN, H.
1957. A comparison of some physiological properties of an inhibitory factor from brain (Factor I) and of gamma-aminobutyric acid and related compounds. J. Physiol. *139*: 79–86.

MAGLADERY, J. W., TEASDALL, R. D., PARK, A. M., and LANGUTH, H. W.
1952. Electrophysiological studies of reflex activity in patients with lesions of the nervous system. I. A comparison of spinal moto-neurone excitability following afferent nerve volleys in normal persons and patients with upper motor neurone lesions. Bull. Johns Hopkins Hosp. *91*: 219–244.

MAGNUS, R.
1909. Zur Regelung der Bewegungen durch das Zentralnervensystem. Pflüger's Arch. ges. Physiol. *130*: 219–252, 253–269.
1910. Zur Regelung der Bewegungen durch das Zentralnervensystem. Pflüger's Arch. ges. Physiol. *134*: 545–583, 584–597.
1924. Körperstellung. Springer, Berlin. 740 pp.
1926. Cameron Prize Lectures on some results of studies in the physiology of posture, parts I and II. Lancet *2*: 531–536, 585–588.

MAGNUS, R., and KLEIJN, A. DE
1912. Die Abhängigkeit des Tonus der Extremitätenmuskeln von der Kopfstellung. Pflüger's Arch. ges. Physiol. *145*: 455–548.
1923. Experimentelle Physiologie des Vestibularapparates bei Säuge-tieren mit Ausschluss des Menschen. *In* Hanb. d. Neurol. d. Ohres *1*: 535. ALEXANDER-MARBURG, ed. Vienna.

MAGOUN, H. W.
1944. Bulbar inhibition and facilitation of motor activity. Science *100*: 549–550.
1950. Caudal and cephalic influences of the brain stem reticular forma-tion. Physiol. Revs. *30*: 459–474.
1953a. An ascending reticular activating system in the brain stem. Harvey Lectures *47*: 53–71. Academic Press, New York.

1953b. (Postural) reticular and striate mechanisms. XIX Intern. Physiol. Congr., Montreal, pp. 87–89.

MAGOUN, H. W., and RHINES, R.
1947. Spasticity: the stretch-reflex and extrapyramidal systems. C. C Thomas, Springfield, Ill. 59 pp.

MARCHAND, L., and SCHIFF, P.
1925. Tumors of corpus callosum inducing epilepsy signs. Encéphale *20*: 512–520.

MARRAZZI, A. S.
1939a. Electrical studies on the pharmacology of autonomic synapses. II. The action of asympathomimetic drug (epinephrine) on sympathetic ganglia. J. Pharmacol. Exp. Therap. *65*: 395–404.
1939b. A self-limiting mechanism in sympathetic homeostatic adjustment. Science *90*: 251–252.
1953. Some indications of cerebral humoral mechanisms. Science *118*: 367–370.

MARRAZZI, A. S., HART, E. R., and COHN, V. H., JR.
1956. Pharmacology of the nervous system. *In* Progress in Neurology and Psychiatry. SPIEGEL, E. A., ed. Grune and Stratton, New York. 606 pp., pp. 565–94.

MARRAZZI, A. S., HART, E. R., PEACOCK, S. M., JR., KOLOVOS, E. R., and ANDY, O. J.
1955. Pharmacology of the nervous system. *In* Progress in Neurology and Psychiatry, SPIEGEL, E. A., ed. Grune and Stratton, New York. 645 pp., pp. 110–132.

MARTINI, E., and TORDA, C.
1938. Der Cholinesterasegehalt des Rückenmarks. Klin. Wochschr. *17*: 98–99.

MATTHEWS, B. H. C.
1929. Specific nerve impulses. J. Physiol. *67*: 169–190.
1931a. The response of a single end organ. J. Physiol. *71*: 64–110.
1931b. The response of a muscle spindle during active contraction of a muscle. J. Physiol. *72*: 153–174.
1933. Nerve endings in mammalian muscle. J. Physiol. *78*: 1–53.

MAYER-GROSS, W.
1951. Experimental psychoses and other mental abnormalities produced by drugs. Brit. Med. J. *2*: 317–321.

MAYFIELD, F. H.
1951. Causalgia. C. C Thomas, Springfield, Ill. 65 pp.

MEDUNA, L. v.
1935. Die Konvulsionstherapie der Schizophrenie. Psychiat. Neurol. Wochschr. *37*: 317–319.

MEIER, R., and BEIN, H. J.
1950. Der Einfluss der Nebennieren auf die Wirkung kreislaufaktiver Substanzen. Helv. Physiol. et Pharmacol. Acta *8*: 436–453.

MELTZER, S. J., and AUER, C. M.
1904. Studies on the "paradoxical" pupil-dilatation caused by adrenalin. Parts I–III. Am. J. Physiol. *11*: 28–51.

MERRITT, H. H., and BRENNER, C.
1941. The effect of acetylcholine on the electrical activity of the cerebral cortex. Trans. Am. Neurol. Assoc. *67*: 152–154.

METTLER, F. A., and CULLER, E.
1934. Action of drugs on the chronic decorticated preparation. J. Pharmacol. Exp. Therap. *52*: 366–377.

MEYERS, R. E., and SPERRY, R. W.
1958. Interhemispheric communication through the corpus callosum. A.M.A. Arch. Neurol. Psychiat. *80*: 298–303.

MILEDI, R.
1959. Acetylcholine sensitivity of partially denervated frog muscle fibres. J. Physiol. *147*: 45P.

MILLER, F. R.
1931. Spinal, bulbar and decerebrate reflexes in the forelimb. J. Physiol. *73*: 1–24.
1934. Reflexes in triceps extensor preparation of forelimb. J. Physiol. *81*: 194–217.
1943. Direct stimulation of the hypoglossal nucleus by acetylcholine in extreme dilutions. Proc. Soc. Exp. Biol. Med. *54*: 285–287.
1949. Effects of eserine and acetylcholine on the respiratory centres and hypoglossal nuclei. Can. J. Research (E) *27*: 374–386.

MILLER, F. R., STAVRAKY, G. W., and WOONTON, G. A.
1940. Effects of eserine, acetylcholine and atropine on the electrocorticogram. J. Neurophysiol. *3*: 131–138.

MINES, G. R.
1913. On dynamic equilibrium in the heart. J. Physiol. *46*: 349–383.

MINZ, B., and DOMINO, E. F.
1953. Effects of epinephrine and norepinephrine on electrically induced seizures. J. Pharmacol. Exp. Therap. *107*: 204–218.

MITCHELL, S. WEIR
1897. Clinical lessons on nervous diseases. Lea Brothers & Co. Philadelphia and New York. 305 pp.

MOLDAVER, J.
1935. Effets immédiats et tardifs, de la déafférentation sur le réflexe contralatéral du chat et du crapaud. Compt. rend. soc. biol. *120*: 514–520.

1936. Contribution à l'étude de la régulation réflexe des mouvements. Arch. intern. méd. exp. *11*: 405–476.

MOMMSEN, J.
1885. Beitrag zur Kenntnis des Muskeltonus. Virchow's Arch. pathol. Anat. u. Physiol. *101*: 22–36.

MONAKOW, C. V.
1914. Die Localization in Grosshirn. J. F. Bergman, Wiesbaden. 1033 pp.

MOORE, B., and OERTEL, H.
1899. A comparative study of reflex action after complete section of the spinal cord in the cervical or upper dorsal region. Am. J. Physiol. *3*: 45–52.

MORITA, S.
1915. Untersuchungen an grosshirnlosen Kaninchen. IV Mitteilung: Quantitative Untersuchungen über die schlafmachende Wirkung von Chloralhydrat und Urethan. Arch. exp. Pathol. Pharmakol. *78*: 223–231.

MORSIER, G. DE., GEORGI, F., and RUTISHAUER, F.
1938. Experimental studies of convulsions provoked by cardiazol in rabbits. Am. J. Psychiat. *94*: 207–208.

MORUZZI, G.
1939. Contribution à l'électrophysiologie du cortex moteur: facilitation, afterdischarge et épilepsie corticales. Arch. intern. physiol. *49*: 33–100.
1946. L'epilessia sperimentale. Nicola Zanichelli, Bologna.
1952. The physiologic mechanisms of epileptic discharge. Acta Psychiat. Neurol. Scand. *27*: 317–328.
1953. (Postural) cerebellar and cerebral mechanisms. XIX Intern. Physiol. Congr., Montreal, pp. 89–99.

MOTT, F. W., and SCHAEFER, E. A.
1890. On movements resulting from faradic excitation of the corpus callosum in monkeys. Brain *13*: 174–177.

MOTT, F. W., and SHERRINGTON, C. S.
1894. Experiments upon the influence of sensory nerves upon movement and nutrition of the limb. Preliminary Communication. Proc. Roy. Soc. London *B 57*: 481–488.

MUNK, H.
1909. Verhalten der niederen Teile des Cerebrospinalsystems nach Ausschaltung höherer Teile. Sitzber. preuss. Akad. Wiss. Physik. math. Kl., Berlin, pp. 1106–1133.

MURRAY, J. G., and THOMPSON, J. W.
1957. The occurrence and function of collateral sprouting in the sympathetic nervous system of the cat. J. Physiol. *135*: 133–162.

NACHMANSOHN, D.
1945. Role of acetylcholine in mechanism of nerve activity. Vitamins and Hormones *3*: 337–377. Academic Press Inc., New York.

NACHMANSOHN, D., and HOFF, E. C.
1944. Effects of dorsal root section on choline esterase concentration in spinal cord of cats. J. Neurophysiol. *7*: 27–36.

NECHELES, H., FRANK, R., KAYE, W., and ROSENMAN, E.
1936. Effect of acetylcholine on the blood flow through the stomach and legs of the rat. Am. J. Physiol. *114*: 695–699.

NECHELES, H., LEVITSKY, P., KOHN, R., MASKIN, M., and FRANK, R.
1936. The vasomotor effect of acetylcholine on the stomach of the dog. Am. J. Physiol. *116*: 330–336.

NICHOL, J., GIRLING, F., JERRARD, W., CLAXTON, E. B., and BURTON, A. C.
1951. Fundamental instability of small blood vessels and critical closing pressures in vascular beds. Am. J. Physiol. *164*: 330–344.

NICHOLLS, J. G.
1956. The electrical properties of denervated skeletal muscle. J. Physiol. *131*: 1–12.

NICKERSON, M., and HOUSE, H. D.
1958. Mechanism of denervation sensitization. Federation Proc. *17*: 398.

NOONAN, T. R., FENN, W. O., and HAEGE, L.
1941. The effects of denervation and of stimulation on exchange of radioactive potassium in muscle. Am. J. Physiol. *132*: 612–621.

NOTKIN, J., and PIKE, F. H.
1931. Some experiments on the effects of caffeine, adrenalin, and bromides upon the susceptibility of experimentally induced convulsions in animals. Am. J. Psychiat. *10*: 771–780.

OBRADOR, A. S.
1942. Corpus callosum and epileptic fits. Bol. lab. estud. med. mexic. *1*: 29.
1947. Hiperexcitabilidad de neurones motoras producida por aislamiento parcial de areas de la corteza cerebral. Rev. clín. españ. *25*: 171–174.

ORLOFF, M. J., WILLIAMS, H. L., and PFEIFFER, C. C.
1949. Timed intravenous infusion of metrazol and strychnine for testing anticonvulsant drugs. Proc. Soc. Exp. Biol. Med. *70*: 254–257.

OWEN, A. G. W., and SHERRINGTON, C. S.
1911. Observations on strychine reversal. J. Physiol. *43*: 232–241.

PACELLA, B. L., KOPELOFF, L. M., and KOPELOFF, N.
 1947. Electroencephalographic studies on incised and excised epileptogenic foci in monkeys. Arch. Neurol. Psychiat. *58*: 693–703.

PARKER, G. H.
 1942. Sensitization of melanophores by nerve cutting. Proc. Nat. Acad. Sci. *28*: 164–170.

PATAKY, I., and PFEIFER, A. K.
 1955. Physiological significance of the acetylcholine blocking agent in the central nervous system. Acta Physiol. Hung. *8*: 221–229.

PATON, W. D. M., and THOMPSON, J. W.
 1953. The mechanism of action of adrenaline on the superior cervical ganglion of the cat. XIX Intern. Physiol. Congr., Montreal, pp. 664–665.

PAVLOV, I. P.
 1928. Lectures on conditioned reflexes: twenty-five years of objective study of the higher nervous activity (behaviour) of animals. Translated from the Russian by W. HORSLEY GANTT. Int. Publishers Co. Inc., New York. 414 pp.
 1941. Lectures on conditioned reflexes. Vol. 2, Conditioned reflexes and psychiatry. Translated and edited by W. HORSLEY GANTT. Int. Publishers Co. Inc., New York. 199 pp.

PEACOCK, S. M., JR.
 1957. Activity of anterior suprasylvian gyrus in response to transcallosal afferent volleys. J. Neurophysiol. *20*: 140–155.

PEELE, T. L.
 1944. Acute and chronic parietal lobe ablations in monkeys. J. Neurophysiol. *7*: 269–286.

PENFIELD, W., and ERICKSON, T.
 1941. Epilepsy and cerebral ablation. C. C Thomas, Springfield, Ill. 623 pp.

PENFIELD, W., and HUMPHREYS, S.
 1940. Epileptogenic lesions of the brain. Arch. Neurol. Psychiat. *43*: 240–261.

PENFIELD, W., and JASPER, H. H.
 1954. Epilepsy and the functional anatomy of the human brain. Little, Brown and Co. Boston. 896 pp.

PENFIELD, W., and RASMUSSEN, T.
 1950. The cerebral cortex of man. Macmillan Co., New York. 248 pp.

PERUZZI, P.
 1947. Contributo alla conoscenza della regolazione neuroumorale degli organi della vita vegetativa; azione difasica, addizione latente subliminale e proporzionalità di effetto di stimoli chimici liminali nel cuore isolato di cavia. Arch. fisiol. *46*: 171–181.

1948. Sulla natura dell 'azione dijasica (motrice e inhibitrice) eserci-
tata dall 'acetilcolina sul cuore isolato dei mammiferi. Boll. soc.
ital. biol. sper. *24*: 797–798.

PETERSON, E. W., MAGOUN, H. W., McCULLOCH, W. S., and LINDSLEY, D. B.
1949. Production of postural tremor. J. Neurophysiol. *12*: 371–384.

PFEIFER, A. K., and PATAKY, I.
1955. Acetylcholine blocking agent in the central nervous system. Acta
Physiol. Hung. *8*: 209–219.

PHILIPEAUX, J. M., and VULPIAN, A.
1863. Note sur une modification physiologique qui se produit dans le
nerf lingual par suite de l'abolition temporaire de la motricité dans
le nerf hypoglosse du même côté. Compt. rend. acad. sci. Paris.
56: 1009–1011.

PHILIPPSON, M.
1905. L'autonomie de la centralisation dans les systèmes nerveux des
animaux. Trav. lab. inst. physiol. Inst. Solvey *7* (pt. ii): 1–208.

PIERCE, F. R., and GREGERSEN, M. I.
1937. Changes in the submaxillary secretory response to pilocarpine
after section of the chorda tympani. Am. J. Physiol. *120*: 246–256.

PIKE, F. H., COOMBS, H. C., and HASTINGS, A. B.
1919. The dependence of the respiratory activity upon conditions in the
central mechanism. Am. J. Physiol. *49*: 125–126.

PIKE, F. H., and ELSBERG, C. A.
1925. Studies on epilepsy: occurrences of clonic convulsive seizures in
animals deprived of cerebral motor cortex. Am. J. Physiol. *72*:
337–342.

PIKE, F. H., ELSBERG, C. A., McCULLOCH, W. S., and RIZZOLO, A.
1929. Some observations on experimentally produced convulsions;
localization of motor mechanisms from which typical clonic
movements of epilepsy arise. Am. J. Psychiat. *9*: 259–283.

PINES, L. J., and MAIMAN, R. M.
1939. Cells of origin of fibers of corpus callosum; experimental and
pathologic observations. Arch. Neurol. Psychiat. *42*: 1076–1082.

PI-SUÑER, J., and FULTON, J. F.
1929. Influence of proprioceptive system upon crossed extensor reflex.
Am. J. Physiol. *88*: 453–467.

POLLOCK, L. J., and DAVIS, L. E.
1922. Experimental convulsions. Research Publs., Assoc. Research
Nervous Mental Disease *7*: 158–175.
1923. Studies in decerebration. I. A method of decerebration. Arch.
Neurol. Psychiat. *10*: 391–398.

1924. Studies in decerebration. II. An acute decerebrate preparation. Arch. Neurol. Psychiat. *12*: 288–293.

1926. Studies in decerebration. III. The labyrinth. Arch. Neurol. Psychiat. *16*: 555–565.

1927a. Studies in decerebration. IV. Integrated reflexes of the brain stem. Arch. Neurol. Psychiat. *17*: 18–23.

1927b. The influence of the cerebellum upon reflex activities of the decerebrate animal. Brain *50*: 277–311.

1929. Muscle tone; extensibility of muscles in decerebrate rigidity. Arch. Neurol. Psychiat. *21*: 19–36.

1930a. Studies in decerebration. V. The tonic activities of a decerebrate animal exclusive of the neck and labyrinthine reflexes. Am. J. Physiol. *92*: 625–629.

1930b. The reflex activities of a decerebrate animal. J. Comp. Neurol. *50*: 377–411.

1931. Studies in decerebration. VI. The effect of deafferentation upon decerebrate rigidity. Am. J. Physiol. *98*: 47–49.

POLLOCK, L. J., FINKELMAN, I., SHERMAN, I. C., and STEINBERG, D. L.
1939. Motor pattern of convulsions produced by various drugs. Elgin Papers *3*: 45–50.

POPE, A., MORRIS, A., JASPER, H., ELLIOTT, K. A. C., and PENFIELD, W.
1947. Histochemical and action potential studies of epileptogenic areas of the cerebral cortex in man and the monkey. Research Publs., Assoc. Research Nervous Mental Disease *26*: 218–233.

PROSSER, C. L.
1940. Acetylcholine and nervous inhibition in the heart of venus mercenaria. Biol. Bull. *78*: 92–102.

PUCINELLI, E.
1933. La regolazione del defluso venoso. I. Influenza dell' adrenalina, acetilcolina, istamina, peptone ed eccitamento del vago. Riv. patol. sper. *10*: 190–233.

RADEMAKER, G. G. I.
1931. Das Stehen. Springer, Berlin. 476 pp.

RADOUCO, C., FROMMEL, E., GOLD, P., GREDER, G., MELKONIAN, D., RADOUCO, S., STRASSBERGER, L., VALETTE, F., and DUCOMMUN, M.
1952a. Étude physio-pathologique de l'action du pentamethylenetetrazol (cardiazol). Arch. inter. pharmacodynamie *92*: 13–38.

1952b. La physiologie de l'épilepsie électrique expérimentale. Arch. inter. pharmacodynamie *92*: 129–162.

RANSON, S. W.
1928. Role of dorsal roots in muscle tonus. Arch. Neurol. Psychiat. *19*: 201–241.

RANSON, S. W., and BILLINGSLEY, P. R.
1916. Studies in vasomotor reflex arcs. II. The conduction of painful afferent impulses in the spinal nerves. Am. J. Physiol. *40*: 571–584.

RANSON, S. W., and DAVENPORT, H. K.
1931. Sensory unmyelinated fibers in spinal nerves. Am. J. Anat. *48*: 331–353.

RANSON, S. W., and HINSEY, J. C.
1929. Crossed extensor reflex in deafferented muscle after transection of brain stem at varying levels. J. Comp. Neurol. *48*: 393–414.
1931. Contralateral flexor reflex, rebound phenomena, co-contraction and reciprocal innervation in spinal and in decerebrate cats. Arch. Neurol. Psychiat. *26*: 247–267.

RAVENTOS, J.
1935. L'action de la nicotine, de l'adrénaline du chlorure de baryum et de l'acétylcholine sur les artères isolées. Compt. rend. soc. biol. *118*: 1016–1018.

RENSHAW, B.
1946. Observations on interaction of nerve impulses in the gray matter and on the nature of central inhibition. Am. J. Physiol. *146*: 443–448.

RETZLAFF, E.
1954. Neurohistological basis for the functioning of paired half-centers. J. Comp. Neurol. *101*: 407–445.
1955. Neurophysiological evidence for simultaneous excitation and inhibition of Mauthner's cells of teleost. Federation Proc. *14*: 120.

RIKER, W. F., and WESCOE, W. C.
1949. Relationship between cholinesterase inhibition and function in neuro-effector system. J. Pharmacol. Exp. Therap. *95*: 515–527.

RINALDI, F., and HIMWICH, H. E.
1955a. Altering responses and actions of atropine and cholinergic drugs. A.M.A. Arch. Neurol. Psychiat. *73*: 387–395.
1955b. Cholinergic mechanism involved in function of mesodiencephalic activating system. A.M.A. Arch. Neurol. Psychiat. *73*: 396–402.
1955c. Drugs affecting psychotic behaviour and the function of the mesodiencephalic activating system. Diseases of Nervous System *16*: 133–141.

RISER, M., GAYRAL, L., and PIGASSOU, J.
1946. Études expérimentales sur l'épilepsie de l'electro-choc. I. L'épilepsie sans cortex. Ann. Medisc. Psychologiques *104*: 270. Abstracted in Epilepsia *3*: 310, 1948.

ROGOWICZ, N.
1885. Ueber pseudomotorische Einwirkung der Ansa Vieussenii auf die Gesichtsmuskeln. Pflüger's Arch. ges. Physiol. *36*: 1–12.

ROSENBLUETH, A.
1950. The transmission of nerve impulses at neuroeffector junctions and peripheral synapses. John Wiley & Sons Inc., New York. 325 pp.

ROSENBLUETH, A., and CANNON, W. B.
1939. Effects of preganglionic denervation on superior cervical ganglion. Am. J. Physiol. *125*: 276–289.

ROSENBLUETH, A., GARCÍA RAMOS, J., and CANNON, W. B.
1945. Los reflejos espinales extensores crusado e ipsilateral. Arch. inst. cardiol. Méx. *15*: 401–454.

ROSENBLUETH, A., LISSÁK, K., and LANARI, A.
1939. An explanation of the five stages of neuromuscular and ganglionic synaptic transmission. Am. J. Physiol. *128*: 31–44.

ROSENBLUETH, A., and LUCO, J. V.
1937. A study of denervated mammalian skeletal muscle. Am. J. Physiol. *120*: 781–797.

RUBIN, M. A., and WALL, C.
1939. Brain potential changes in man induced by metrazol. J. Neurol. Psychiat. *2*: 107–114.

RUCH, T. C.
1955. Neural basis of somatic sensation: a textbook of physiology. FULTON, J. F., ed. 17th ed. W. B. Saunders Co., Philadelphia and London, 1275 pp.

SACHS, A.
1937. Die Wirkung der Cholinkörper auf den Vorhofstreifen. Cardiologia *1*: 74–87.

SACKS, J., and GLASER, N. M.
1941. Changes in susceptibility to the convulsant action of metrazol. J. Pharmacol. Exp. Therap. *73*: 289–295.

SAKAMOTO, M.
1935. Ueber die Einwirkung von verschiedenen Gefässmitteln auf die Hodengefässe des Kaninchens. Folia Endocrinol. Japon. *11*: 40–41.

SAL Y ROSAS, F.
1943. Diferencias de la susceptibilidad convulsiva experimental Peru y en los paises Europeos. (Diagnosis differences of susceptibility to metrazol convulsions in Peru and in European countries. Experimental study in epileptics and non-epileptics.) Rev. mex. psiquiat. neurol. y med. leg. *10*: 3–16.

SARGANT, W.
1957. Battle for the mind. W. Heinemann Ltd., London. 263 pp.

SAUERBRUCH, F.
1913. Experimentelle studien über die Entstehung der Epilepsie. Verhandl. deut. Ges. Chir. *42*: 144–149.

SCARFF, J. E., and POOL, J. L.
1946. Factors causing massive spasm following transection of the cord in man. J. Neurosurg. *3*: 285–293.

SCHLICHTER, W., BRISTOW, M. E., SCHULTZ, S., and HENDERSON, A. L.
1956. Seizures occurring during intensive chlorpromazine therapy. Can. Med. Assoc. J. *74*: 365–366.

SCHREINER, L. H., LINDSLEY, D. B., and MAGOUN, H. W.
1949. Role of brain stem facilitatory systems in maintenance of spasticity. J. Neurophysiol. *12*: 207–216.

SEGUIN, J. J.
1956. The effects of convulsant and anticonvulsant agents on partially isolated regions of the central nervous system. Ph.D. Thesis, University of Western Ontario, London, Canada. 177 pp.

SEGUIN, J. J., FRETZ, N. A., MANAX, S. J., and STAVRAKY, G. W.
1961. Effect of section of the corpus callosum on pentylenetetrazol convulsions. Arch. Neurol. (In Press)

SEGUIN, J. J., FRETZ, N. A., and STAVRAKY, G. W.
1957. Convulsive activity of corpus-callotomized and semidecerebrated white rats. Proc. Can. Physiol. Soc. *21*: 54. Rev. can. biol. *16*: 514.

SEGUIN, J. J., and STAVRAKY, G. W.
1952. The effect of sodium pentobarbital on neurones sensitized by partial isolation. Proc. Can. Physiol. Soc. *16*: 58–59. Rev. can. biol. (1954) *13*: 86.
1954. Effect of barbiturates on chronic frontal-lobectomized or semidecerebrate cats. Federation Proc. *13*: 403.
1957. The effects of barbiturates on partially isolated regions of the central nervous system. Can. J. Biochem. Physiol. *35*: 667–680.

SELETZKY, W., and GILULA, J.
1928. Zur Frage der Funktionen des Balkens bei Tieren. Arch. Psychiat. Nervenkrankh. *86*: 57–73.

SETSCHENOW, I. M.
1863. Physiologische Studien über die Hemmungsmechanismen für die Reflexthätigkeit des Rückenmarks im Gehirne des Frosches. August Hirschwald, Berlin. 51 pp.
1868. Ueber die elektrische und chemische Reizung der sensiblen Rückenmarksnerven des Frosches. Leuschner and Lubensky, Graz. 69 pp.
1875. Notiz, die reflexhemmenden Mechanismen betreffend. Pflüger's Arch. ges. Physiol. *10*: 163–164.

SHERRINGTON, C. S.

1893. Experiments in examination of the peripheral distribution of the fibres of the posterior roots of some spinal nerves. Phil. Trans. Roy. Soc. London *184*, Ser. B, 641–763.

1894. On the anatomical constitution of nerves of skeletal muscles; with remarks on recurrent fibres in the ventral spinal nerve root. J. Physiol. *17*: 211–258.

1898a. Experiments in examination of the peripheral distribution of the fibres of the posterior roots of some spinal nerves. Part II. Phil. Trans. Roy. Soc. London *190*, Ser. B, 45–186.

1898b. Decerebrate rigidity and reflex coordination of movements. J. Physiol. *22*: 319–332.

1900. On the innervation of antagonistic muscles. Sixth Note. Proc. Roy. Soc. London *66*: 66–67.

1905a. On reciprocal innervation of antagonistic muscles. Seventh Note. Proc. Roy. Soc. London *B 76*: 160–163.

1905b. On reciprocal innervation of antagonistic muscles. Eighth Note. Proc. Roy. Soc. London *B 76*: 269–297.

1906. The integrative action of the nervous system. Constable and Co., London. 411 pp.

1907. Strychnine and reflex inhibition of skeletal muscle. J. Physiol. *36*: 185–204.

1909. On plastic tonus and proprioceptive reflexes. Quart. J. Exp. Physiol. *2*: 109–156.

1910. Flexion-reflex of the limb, crossed extension-reflex, and reflex stepping and standing. J. Physiol. *40*: 28–121.

1915. Postural activity of muscle and nerve. Brain. *38*: 191–234.

SHERRINGTON, C. S., and SOWTON, S. C. M.

1911a. Reversal of the reflex effect of an afferent nerve by altering the character of the electrical stimulus applied. Proc. Roy. Soc. London, *B 83*: 435–446.

1911b. On reflex inhibition of the knee flexor. Proc. Roy. Soc. London, *B 84*: 201–214.

1911c. Chloroform and reversal of reflex effect. J. Physiol. *42*: 383–388.

SIGGS, E., OCHS, S., and GERARD, R. W.

1955. Effects of the medullary hormones on the somatic nervous system in the cat. Am. J. Physiol. *183*: 419–426.

SIMEONE, F. A.

1938. The effect of previous stimulation on the responsiveness of the cat's nictitating membrane sensitized by denervation. Am. J. Physiol. *122*: 650–658.

SIMEONE, F. A., and MAES, J. P.

1939. Sensitization of the submaxillary gland by sympathetic denervation. Am. J. Physiol. *125*: 674–679.

SJÖSTRAND, T.
1937. Potential changes in the cerebral cortex of the rabbit arising from cellular activity and the transmission of impulses in the white matter. J. Physiol. *90*: 41–43P.

SMITH, C. G., METTLER, F. A., and CULLER, E. A.
1940. The phasic responses to cortical stimulation. J. Neurophysiol. *3*: 182–187.

SMITH, D. C.
1941. The effect of denervation upon the response to adrenalin in isolated fish scale melanophore. Am. J. Physiol. *132*: 245–248.

SMITH, K. U., and AKELAITIS, A. J.
1942. Studies on corpus callosum. Laterality in behaviour and bilateral motor organization in man before and after section of corpus callosum. Arch. Neurol. Psychiat. *47*: 519–543.

SNELL, R. S., and MCINTYRE, N.
1955. Effect of denervation on the histochemical appearance of cholinesterase at the myoneural junction. Nature *176*: 884–885.

SOCIN, C., and LEEUWEN, W. STORM VAN
1914. Ueber den Einfluss der Kopfstellung auf phasische Extremitätenreflexe. Pflüger's Arch. ges. Physiol. *159*: 251–275.

SOLANDT, D. Y., and MAGLADERY, J. W.
1942. A comparison of effects of upper and lower motor neurone lesions on skeletal muscle. J. Neurophysiol. *5*: 373–380.

SOLANDT, D. Y., PARTRIDGE, R. C., and HUNTER, J.
1943. The effect of skeletal fixation on skeletal muscle. J. Neurophysiol. *6*: 17–22.

SPADOLINI, I.
1917. Le azioni antagonistiche nei sistemi autonomi, rivista critica e riceiche sperimentale, Arch. fisiol. *15*: 1–167.
1948. Researches on diphasic action of chemical mediators and contribution to humoral regulation of autonomic effectors. Arch. intern. physiol. *55*: 317–333.

SPADOLINI, I., and DOMINO, G.
1940. La duplice azione dell'acetilcolina sul cuore isolato di cavia: Considerazioni sulla recettività cellulare negli organi della vita vegetativa. Arch. fisiol. *40*: 147–171.

SPARKS, M. I.
1927. Experimental studies of epileptiform convulsions. Arch. intern. pharmacodynamie *33*: 460–481.

SPIEGEL, E. A.
1937. Comparative study of the thalamic cerebral and cerebellar potentials. Am. J. Physiol. *118*: 569–579.

1960. Neural mechanisms in extrapyramidal diseases and their treatment (Neurosurgical Panel). *In* International Symposium on the Extrapyramidal System and Neuroleptics, Montreal. (In Press)

SPIEGEL, E. A., and DÉMÉTRIADES, TH. D.
1925. Die zentrale Kompensation des Labyrinthverlustes. Pflüger's Arch. ges. Physiol. *210*: 215–222.

SPIEGEL, E. A., KLETZKIN, M., SZEKELY, E. G., and WYCIS, H. T.
1954. Role of hypothalamic mechanisms in thalamic pain. Neurology *4*: 739–751.

SPIEGEL, E. A., and SCALA, N. P.
1943. Response of labyrinthine apparatus to electrical stimulation; site of action; faradic stimulation; inverse effects of anodic, and cathodic stimulation. Arch. Otolaryngol., Chicago *38*: 131–138.

SPIEGEL, E. A., and SZEKELY, E. G.
1955. Supersensitivity of the sensory cortex following partial deafferentation. Electroencephalog. and Clin. Neurophysiol. *7*: 375–381.

SPIEGEL, E. A., and WYCIS, H. T.
1953. Mesencephalotomy in treatment of "intractable" facial pain. A.M.A. Arch. Neurol. Psychiat. *69*: 1–13.

SPIEGEL, E. A., WYCIS, H. T., BAIRD, H. W. 3RD, and SZEKELY, E. G.
1959. Physiopathologic observations on the basal ganglia. First International Congress of Neurological Sciences, Brussels, 1957. Joint Proceedings and Round Table Conference. Pergamon Press, New York. Vol. *5*: 118–122.

SPRAGUE, J. M., SCHREINER, L. H., LINDSLEY, D. B., and MAGOUN, H. W.
1948. Reticulo-spinal influences on stretch reflexes. J. Neurophysiol. *11*: 501–507.

SPRONG, W. L.
1929. Study of reflexes in deafferented leg of cat and their relation to tonus. Bull. Johns Hopkins Hosp. *45*: 371–395.

STAMM, J. S., and SPERRY, R. W.
1957. Function of corpus callosum in contralateral transfer of somesthetic discrimination in cats. J. Comp. and Physiol. Psychol. *50*: 138–143.

STARZL, T. E., TAYLOR, C. W., and MAGOUN, H. W.
1951. Collateral afferent excitation of reticular formation of brain stem. J. Neurophysiol. *14*: 479–496.

STAVRAKY, G. W.
1934. Reversal effect of chorda tympani stimulation. J. Pharmacol. Exp. Therap. *50*: 79–87.

1942. The mechanism of the synergistic action of pilocarpine and adrenaline on salivary secretion. Rev. can. biol. *1*: 64–71.

1943. Some aspects of the effects of intravenous injections of acetylcholine on the central nervous system. Trans. Roy. Soc. Can. 3rd series, Sect. V. *37*: 127–139.

1947. The action of adrenaline on spinal neurones sensitized by partial isolation. Am. J. Physiol. *150*: 37–45.

1960. Effects of partial denervation on spinal neurones and their possible relation to Parkinsonism. *In* International Symposium on the Extrapyramidal System and Neuroleptics, Montreal. (In Press)

STEIN, M. H., and WORTIS, H.
1941a. Tabes dorsalis; evaluation of sensory findings. Arch. Neurol. Psychiat. *46*: 471–476.

STEIN, M. H., WORTIS, H., and JOLLIFFE, N.
1941b. Peripheral neuropathy: evaluation of sensory findings. Arch. Neurol. Psychiat. *46*: 464–470.

STEINKE, C. R.
1918. Surgery of posterior spinal roots. Surg. Gynecol. Obstet. *27*: 55–57.

STELLA, G.
1944a. Sul meccanismo della rigidità da decerebrazione in arti deafferentati. Atti soc. med. chir. Padova. *23*: 5–16.

1944b. Influenza del cervelletto sulla rigidità da cerebrazione. Atti soc. med. chir. Padova *23*: 17–21.

1944c. Nuove osservazioni sugli effetti della stimolazione del paleocerebello dell'animale decerebrato. Atti soc. med. chir. Padova *23*: 22–24.

1946. Nuove asservazione sull'attività riflessa tonica del cervelletto nello animale decerebrato. Boll. soc. ital. biol. sper. *22*: 78–80.

STRAUSS, H., and LANDIS, C.
1938. Metrazol convulsions and their relation to epileptic attack. Proc. Soc. Exp. Biol. Med. *38*: 369–370.

STRAUSS, H., LANDIS, C., and HUNT, W. A.
1939. Metrazol seizure and its significance for pathophysiology of epileptic attack. J. Nervous Mental Disease *90*: 439–452.

STRAUSS, H., RAHM, E. W., JR., and BARRERA, S. E.
1939. Electroencephalographic studies on relatives of epileptics. Proc. Soc. Exp. Biol. Med. *42*: 207–212.

STRÖMBLAD, R.
1955. Acetylcholine inactivation and acetylcholine sensitivity in denervated salivary glands. Acta Physiol. Scand. *34*: 38–58.

SWANK, R. L.
1949. Synchronization of spontaneous electrical activity of cerebrum by barbiturate narcosis. J. Neurophysiol. *12*: 161–172.

TEASDALL, R. D.
1950. A study of the responses of partially denervated (deafferented) neurones of the central nervous system to pyramidal impulses and to reflex and chemical stimulation. Ph.D. Thesis, University of Western Ontario, London, Canada. 395 pp.

TEASDALL, R. D., and MAGLADERY, J. W.
1956. Changes in spinal reflex patterns following hemisection. Trans. Am. Neurol. Assoc. *81*: 152–153.

TEASDALL, R. D., MAGLADERY, J. W., and RAMEY, E. H.
1958. Changes in reflex patterns following spinal cord hemisection in cats. Bull. Johns Hopkins Hosp. *103*: 223–235.

TEASDALL, R. D., PARK, A. M., LANGUTH, H. W., and MAGLADERY, J. W.
1952. Electrophysiological studies of reflex activity in patients with lesions of the nervous system. II. Disclosure of normally suppressed monosynaptic reflex discharges of spinal motoneurones by lesions of lower brainstem and spinal cord. Bull. Johns Hopkins Hosp. *91*: 245–256.

TEASDALL, R. D., and STAVRAKY, G. W.
1950. Effect of section of the corpus callosum on experimental convulsions. Federation Proc. *9*: 124–125.
1953. Responses of deafferented spinal neurones to corticospinal impulses. J. Neurophysiol. *16*: 367–375.
1955. Tonic adaptations in deafferented limbs of the cat. Can. J. Biochem. Physiol. *33*: 139–155.

TEN CATE, J., and SWIJGMAN, D. W.
1945. Localization de l'origine des convulsions produites par le cardiazol et la coramine. Arch. intern. pharmacodynamie. *70*: 293–306.

TERZIAN, H., and TERZUOLO, C.
1951. Sul meccanismo della rigidità da decerebrazione in arti anteriori cronicamente deafferentati. Boll. soc. ital. biol. sper. *27*: 1317–1318.
1954. Le componenti automatiche e riflesse del tono posturale. Arch. fisiol. *54*: 37–61.

TERZUOLO, C., and TERZIAN, H.
1951. Riflessi di magnus e tono posturale in arti sottoposti a deafferentazione acuta o cronica. Boll. soc. ital. biol. sper. *27*: 1319–1320.
1953. Cerebellar increase of postural tonus after deafferentation and labyrinthectomy. J. Neurophysiol. *16*: 551–561.

THESLEFF, S.
1960. Effects of motor innervation on the chemical sensitivity of skeletal muscle. Physiol. Revs. *40*: 734–752.

THIELE, F. H.
1905. On the efferent relationship of the optic thalamus and Deiter's nucleus to the spinal cord, with special reference to the cerebellar influx of Dr. Hughlings Jackson and the genesis of the decerebrate rigidity of Ord and Sherrington. J. Physiol. *32*: 358–384.

TOMAN, J. E. P., and GOODMAN, L. S.
1947. Conditions modifying convulsions in animals. Research Publs., Assoc. Research Nervous Mental Disease *26*: 141–162.
1948. Anticonvulsants. Physiol. Revs. *28*: 409–432.

TOMAN, J. E. P., SWINYARD, E. A., and GOODMAN, L. S.
1946. Properties of maximal seizures and their alteration by anticonvulsant drugs and other agents. J. Neurophysiol. *9*: 231–239.

TOWER, D. B., and ELLIOTT, K. A. C.
1952. Activity of the acetylcholine system in the human epileptogenic foci. J. Appl. Physiol. *4*: 669–676.

TOWER, D. B., and McEACHERN, D.
1949a. Acetylcholine and neuronal activity. I. Cholinesterase patterns and acetylcholine in the cerebrospinal fluids of patients with craniocerebral trauma. Can. J. Research, Sect. E. *27*: 105–119.
1949b. Acetylcholine and neuronal activity. II. Acetylcholine and cholinesterase activity in the cerebrospinal fluids of patients with epilepsy. Can. J. Research, Sect. E. *27*: 120–131.

TOWER, S. S.
1937. Function and structure in the chronically isolated lumbo-sacral spinal cord of the dog. J. Comp. Neurol. *67*: 109–131.

TOWER, S., BODIAN, D., and HOWE, H.
1941. Isolation of intrinsic and motor mechanism of the monkey's spinal cord. J. Neurophysiol. *4*: 388–397.

TRENDELENBURG, W.
1906. Ueber die Bewegung der Vögel nach Durchschneidung hinterer Rückenmarkswurzeln: ein Beitrag zur Physiologie des Zentralnervensystems der Vögel (nach Untersuchungen an Columbia Domestica) Arch. Anat. u. Physiol., Physiol. Abt. pp. 1–126.
1910. Untersuchungen über reizlose vorübergehende Ausschaltung am Zentralnervensystem. II. Mitteilung. Zur Lehre von den bulbären und spinalen Atmungs—und Gefässzentren. Pflüger's Arch. ges. Physiol. *135*: 469–505.

TWAROG, B. M.
1954. Responses of malluscan smooth muscle to acetylcholine and 5-hydroxytryptamine. J. Cellular Comp. Physiol. *44*: 141–163.

UCHINO, K.
1935. Ueber die Einwirkung von verschiedenen Gefässmitteln auf die Ovarialgefässe des Rindes. Folia Endocrinol. Japon. *11*: 54.

188 *References*

UEXKULL, J. V.
    1904. Studien über den Tonus. Z. Biol. *46*: 1–37.

UYEMATSU, S., and COBB, S.
    1922. Preliminary report on experimental convulsions; convulsions produced by administration of chemical substances. Arch. Neurol. Psychiat. *7*: 660–661.

VERWORN, M.
    1900. Zur Kenntniss der physiologischen Wirkungen des Strychnins. Arch. Anat. u. Physiol., Physiol. Abt. pp. 385–414.

VOGT, M.
    1954. Concentration of sympathin in different parts of central nervous system under normal conditions and after administration of drugs. J. Physiol. *123*: 451–481.

    1957. Sympathomimetic amines in the central nervous system: normal distribution and changes produced by drugs. Brit. Med. Bull. *13*: 166–171.

WAGENEN, W. P. VAN, and HERREN, R. Y.
    1940. Surgical division of commissural pathways in corpus callosum: relation to spread of epileptic attack. Arch. Neurol. Psychiat. *44*: 740–759.

WALKER, A. E.
    1942a. Relief of pain by mesencephalic tractotomy. Arch. Neurol. Psychiat. *48*: 865–880.
    1942b. Somatotopic localization of spinothalamic and secondary trigeminal tracts in mesencephalon. Arch. Neurol. Psychiat. *48*: 884–889.
    1949. Post-traumatic epilepsy. C. C Thomas, Springfield, Ill. 86 pp.
    1960. Neural mechanisms in extrapyramidal diseases and their treatment (Neurosurgical Panel). *In* International Symposium on the Extrapyramidal System and Neuroleptics, held at University of Montreal. (In Press)

WALSH, E. G.
    1957. Physiology of the nervous system. Longmans, Green and Co. London. 563 pp.

WALSHE, F. M. R.
    1928. Quoted by BREMER. *In* De l'exagération des réflexes, consecutives à la section des racines postérieures, Ann. physiol. physicochim. biol. *4*: 750–752.

WARD, A., JR., and JENKNER, F. L.
    1953. Bulbar reticular formation and tremor. Trans. Am. Neurol. Assoc. *78*: 36–37.

WARD, A. A., JR., and KENNARD, M. A.
  1942. Effect of cholinergic drugs on recovery of function following lesions of the central nervous system in monkeys. Yale J. Biol. Med. *15*: 189–228.

WARD, A. A., JR., McCULLOCH, W. S., and MAGOUN, H. W.
  1948. Production of an alternating tremor at rest in monkeys. J. Neurophysiol. *11*: 317–330.

WARD, J. W., and CLARK, S. L.
  1938. Convulsions produced by electrical stimulation of the cerebral cortex of unanesthetized cats. Arch. Neurol. Psychiat. *39*: 1213–1227.

WEDDELL, G., GUTTMANN, L., and GUTMANN, E.
  1941. Local extension of nerve fibers into denervated areas of skin. J. Neurol. Psychiat. *4*: 206–225.

WEDENSKY, N. E.
  1903. Die Erregung, Hemmung and Narkose. Pflüger's Arch. ges. Physiol. *100*: 1–144.

WEED, L. H.
  1914. Observations upon decerebrate rigidity. J. Physiol. *48*: 205–227.

WÉGRIA, RENÉ
  1951. Pharmacology of coronary circulation. Pharmacol. Revs. *3*: 197–246.

WELSH, J. H.
  1948. Concerning mode of action of acetylcholine. Bull. Johns Hopkins Hosp. *83*: 568–586.
  1955. Neurohormones. *In* PINCUS, G., and THIAMANN, K. V., eds., The Hormones, Physiology, Chemistry and Applications. Vol. III. Academic Press Inc., N.Y. 1012 pp., pp. 97–151.

WIEDEMAN, M.P.
  1954. Reactivity of arterioles following denervation of subcutaneous areas of the bat wing. Am. J. Physiol. *177*: 208–314.
  1955. Effect of denervation on diameter and reactivity of arteries in the bat wing. Circulation Research. *3*: 618–622.

WIKLER, A.
  1950. Sites and mechanisms of action of morphine and related drugs in the central nervous system. Pharmacol. Revs. *2*: 435–506.

WIKLER, A., and FRANK, K.
  1948. Effects of electroshock convulsions on chronic decorticated cats. Proc. Soc. Exp. Biol. Med. *67*: 464–468.

WILLS, J. H.
  1942. Sensitization of the submaxillary gland to acetylcholine by section of the chorda tympani. Am. J. Physiol. *135*: 523–525.

190 *References*

WILSON, S. A. K.
1908. A contribution to the study of apraxia with a review of the literature. Brain *31*: 164–216.

WOLFF, H. G., and CATTELL, McK.
1937. On the mechanism of hypersensitivity produced by denervation. Am. J. Physiol. *119*: 422–423.

WORTIS, S. B., COOMBS, H. C., and PIKE, E. H.
1931. Monobromated camphor: standardized convulsant. Arch. Neurol. Psychiat. *26*: 156–161.

WORZNIAK, J., and GESELL, R.
1938. On central action of acetylcholine. Am. J. Physiol. *123*: 222.

WYSS, O. A. M.
1940. Ein weiterer Beitrag zur Kenntnis vom Mechanismus der vagalen Atmungssteuerung. Pflüger's Arch. ges. Physiol. *243*: 457–467.
1944. Le rôle physiologique de la fréquence des influx afferents. Compt. rend soc. phys. et hist. nat. Genève. *61*: 63–66.
1947a. Reflex reversal as determined by the frequency of afferent stimulation. Arch. néerl. physiol. *28*: 444–450.
1947b. Respiratory effects from stimulation of the afferent vagus nerve in the monkey. J. Neurophysiol. *10*: 315–320.

YAKOLEV, P. I.
1937. Neurologic mechanism concerned in epileptic seizures. Arch. Neurol. Psychiat. *37*: 523–554.

ZADOR, J.
1930. Meskalinwirkung auf das Phantomglied Beitrag zur neurophysiologischen Betrachtung der Wahrnehmung und Vorstellung. Montasschr. Psychiat. Neurol. *77*: 71–99.

ZELLER, E. A., and BISSEGGER, A.
1943. Ueber die Cholin-Estèrase des Gehirns und der Erythrocyten. Helv. Chim. Acta *26*: 1619–1630.

ZISKIND, E., and BERCEL, N. A.
1947. Preconvulsive paroxysmal electroencephalographic changes after metrazol injections. Research Publs., Assoc. Research Nervous Mental Disease *26*: 487–501.

www.ingramcontent.com/pod-product-compliance
Lightning Source LLC
Chambersburg PA
CBHW030507210326
41597CB00013B/824